SALT
The Complete Reference

The Author

Param H Dave is experienced person in Solar Salt industry. He is working in TATA Chemicals Limited, India. TATA Chemicals Limited is world's leading Soda Ash and Salt manufacturer. At Tata, author is In-charge in Solar Salt production. He is responsible for entire brine field to produce Solar Salt up to 2.2 million metric ton and equivalent brine for 1.0 million metric ton of Vacuum Salt. Author is having rich experience in salt works design, development, legal aspects of solar salt, environment aspects and Salt production procedure with scientific manner.

In early career he worked as an IT professional with GPS mapping and Machine automation in different industries. Author is having also experience in Environment and Sustainable development. He got certification from Harvard Business School on Process Improvement and Crisis Management

While working in salt industry he feels that there is no any reference book or hand book so author decided to prepare a book on SALT.

SALT
The Complete Reference

Param H Dave

2018
Daya Publishing House®
A Division of
Astral International Pvt. Ltd.
New Delhi – 110 002

ISBN: 9789388173070 (Int. Edition)

Publisher's Note:

Every possible effort has been made to ensure that the information contained in this book is accurate at the time of going to press, and the publisher and author cannot accept responsibility for any errors or omissions, however caused. No responsibility for loss or damage occasioned to any person acting, or refraining from action, as a result of the material in this publication can be accepted by the editor, the publisher or the author. The Publisher is not associated with any product or vendor mentioned in the book. The contents of this work are intended to further general scientific research, understanding and discussion only. Readers should consult with a specialist where appropriate.

Every effort has been made to trace the owners of copyright material used in this book, if any. The author and the publisher will be grateful for any omission brought to their notice for acknowledgement in the future editions of the book.

Published by : **Daya Publishing House®**
A Division of
Astral International Pvt. Ltd.
– ISO 9001:2015 Certified Company –
4736/23, Ansari Road, Darya Ganj
New Delhi-110 002
Ph. 011-43549197, 23278134
E-mail: info@astralint.com
Website: www.astralint.com

Preface

SALT (NaCl) is a precious gift of nature to mankind. In my work in the Salt industry I always looking for some literature, documents and scientific papers for better outcome in Solar Salt production. Unfortunately there is no any comprehensive documents or book available in Solar Salt industry. Even the most experienced persons who are engaged in Solar Salt industry have very limited knowledge on the subject. So, during my operation at Tata Chemicals Limited, Mithapur, India and Dev Salt Pvt. Ltd. India, I always seeking for literature about Solar Salt industry for my knowledge and field key. Therefore I present this book with the help of some scientific papers, books and my field experiences for ready reference of Solar Salt production operations. I found considerable difficulties in gathering the information necessary for this book. Solar Salt operations are vary from area to area throughout the world, so there may be possible variance in some aspects. The work, however is not the ultimate guideline in the subject but it gives in general outline for the process and can be useful as a field key of Solar Salt operations.

I hope that the present book will be useful for the wide range of readers who are engaged in Solar Salt industry.

Param H. Dave

"Samudram"

Omkar Society. 150′ Ring Road.

Near Ayodhya Nagar. Behind Gokul Mathura apt.

Rajkot 360006

Gujarat (INDIA)

paramhdave@gmail.com

+91 9824999992

Acknowledgement

Before presenting this book to the readers, I have to acknowledge the opportunities and help given to me by my family, my colleagues and my seniors and specially Tata Chemicals Limited. I acknowledge the Tata Chemicals Limited to give me an opportunity to serve its prestigious organization. During my service at Tata Chemicals Limited, Mithapur, I got huge knowledge about the industry, continuous guidance and support from senior leadership team. I am also thankful to Dev Salt Private Limited where I started my journey in Salt industry.

Param H. Dave

Contents

1 Introduction

Common salt (Sodium Chloride - NaCl) is an important mineral required for the body metabolism. However it has many more applications for non-edible purposes. About 85% of the Salt produced in the world is used for non-edible purposes: mainly for chemical industries, deicing, water softening, tanneries, fish-curing etc. Soda Ash, Caustic soda and chlorine produced from salt have immense downstream application like in manufacture of plastic, paints, paper, textiles, soaps, detergents etc. Because of its wide industrial application Common Salt is also called 'Mother of Chemical industries'.

A major portion of the salt used today is Solar Salt. Most of mined salt in the world is the from ancient sea beds where salt was deposited and later covered with deposits of clay, sand or rock. Since these deposits were formed naturally, they all contain various level of impurities. Mined salt, normally called rock salt requires extensive washing processes to refine to a level suitable for human use. Solar Salt produced from sea water can be controlled to produce salt that is virtually as pure and clean as vacuum evaporated salt if it produced in proper scientific manner with care. Vacuum salt is about 99.9% pure NaCl, Solar Salt is routinely produced by the world's major salt plants at purity of 99.7% pure NaCl. The operating procedures, the Theories and Practices to achieve that level of purity with high in quantity will be explained in this book, all other aspects related to Solar Salt will be discussed.

History of Salt

Salt, Sodium Chloride (NaCl) has probably been with us from the beginnings of geologic time, and has probably always been necessary either directly or indirectly through all stages of the evolution of living things. According to geophysicists, it is controversial whether or not the first ocean on the earth after its formation approached the salinity of the present oceans. Even if not, the first oceans undoubtedly contained traces of salt, and our unicellular ancestors first appeared and thrived in this salty marine environment.

In the process of evolution, these unicellular ancestors became multi cellular, and some left their salty marine environment but still required salt. Our herbivorous ancestors used salt licks, and our carnivorous once obtained their salt from the flesh and blood of their prey. This demands - this necessity of all living things for salt in one form or another continuous today.

Our bodies contain relatively large quantities of salt, as we may infer from the taste of "blood, sweat, and tears". Salt water, self – contained, bathes all animal matter, and it might be said we are carrying around with us and living in a portable marine environment such as was used by our unicellular ancestors.

Considered thus, it is easy to see why the earliest records of salt and its used fade backward until lost in the mists of antiquity. Certain it is, that earliest recorded history tells of salt and its uses in a manner so common place that salt was obviously old long before. Salt seems to have been used almost universally, in all times and places, and to have been pursued by man by all means necessary, including war, fraud, ingenuity, and herculean effort.

According to Pliny, "We may conclude then, by Hercules that the higher enjoyments of life could not exist without the use of salt: indeed, so highly necessary is this substance to mankind that the pleasure of the mind, even, can be expressed by no better term than the word SALT such being the name given to all effusions of wit. All the amenities, in fact, of life, supreme hilarity, and relaxation from toil, can find no word in our language to characterize them better than this.

Solar Salt – An Introduction

Production of salt by solar evaporation of sea water has been practiced for centuries. The ancient Greeks, Romans, Egyptians, Indian, were all familiar with this method. The idea probably originated from seeing salt crusts around pools of sea water trapped by receding high tide, and evaporating under summer sun and wind. Observation must also have been made on the salt crusts left on rocks and brush by evaporation of fine sea spray. It would follow in natural sequence that sea water should be led into shallow artificial ponds and crude salt obtained Through evaporation by sun and wind.

This early solar salt was a very improper product. The evaporating stages were not separated, and each brine batch was reduced as nearly to dryness as possible. The product was dump mass with a very bitter taste, caused by high content of bittern salts, and containing only about 65% NaCl, with some 28% hydrated magnesium salts, 4% $CaSO_4 2H_2O$ (Gypsum), plus small amounts of most of the other compounds present in ocean water.

For centuries this method was not improved. It was gradually found that the salt had a better taste if removed from pond before all the brine was evaporated, thus eliminating most of the bittern salts, which distorted the true salt flavor. It was further noticed that the first deposition to appear during early evaporation was very unlike salt. So, after evaporation had proceeded until small salt-tasting crystals formed on the surface, the brine was transferred to second pond where evaporation continued until but a small amount of brine remained. This was drained off, and the salts raked into piles or stacks along the edges of the pond for further draining until ready to use. Thus the quantity and taste of the salt was gradually much improved. Compared to modern high quality, it was still a rather impure product, but the need for separate steps in the process had become recognized.

Solar salt production was of great importance to any country bordered by the sea or gifted by salt water lake. The raw material was inexhaustible, and if climate permitted, the solar salt works system was usually the cheapest way for a country to obtain at least enough salt for its own use. The solar salt works could be large or small.

2

Design and Layout of Solar Salt Works

The success of an ideal salt works depends mainly on the optimum design and layout. Maximum yield and higher purity of salt can be achieved by proper layout. Based on the gross yearly and seasonal evaporation rate, percolation losses, initial density of brine or seawater, number of available days, evaporation rate of different density brines and expected production. Empirical relationship between these parameters are worked out. Evaporation rate of different density brine is correlated with fresh water evaporation rate in a standard evaporimeter to obtain the appropriate surface area required for each pond holding a specific density range. Percolation losses are computed by taking 15 cm depth of saturated brine as the basis for calculations. Viscosity, density and the depth of brine are taken into consideration to obtain net losses of brine due to percolation in an earth-lined pond. The effect of initial density of incoming sea water on the ratio of crystallizer area to the remaining area is also shown in the calculations.

These computations form a guideline whenever a new salt works is to be laid out or improvement is sought in an existing salt works.

The present process of salt manufacture by solar evaporation is based on the experience gathered through centuries and on the background of scientific knowledge of the chemistry involved in the separation of different constituents of seawater or subsoil brines. With the characteristic constancy of composition of the raw materials, the process of salt manufacture involves the same general principles. However, the total area requirement, the ratio of the area of other evaporating ponds to that of crystallizer and operational controls very with climatological conditions and soil nature.

Metrological factors such as maximum/minimum temperatures, relative humidity, wind velocity, incident solar radiation, rain fall and total number of clear days have been considered so far to estimate the exact evaporation rate of brine. Most of the time these extremely elaborate formulae fail to give ultimate correct values in spite of taking every factor into consideration. What is needed is the exact gross evaporation rate of fresh water from a standard evaporimeter which can be used for computing brine evaporation rate.

Percolation of brine through a soil bed at all evaporating stages is an elusive but important factor that has been neglected so far because of its complexity. Though it is impracticable to measure permeability of the entire area under consideration, percolation values computed from the particle size analysis of representative soil samples will definitely lead to a fairly accurate approximation of brine lost through the soil.

This chapter summarizes the importance of these two factors in designing a solar salt works with the help of empirical formulae for calculating area of any pond of a given evaporation rate, percolation and nominal design production level.

Rate of Brine Evaporation

Fairly accurate values of the rate of brine evaporation can be calculated if gross fresh water evaporation rate is known from standard evaporimeter. The calculations in subsequent paragraphs are based on the values obtained from a standard U.S. 'A' pan evaporimeter of 1220 mm diameter and 250 mm water depth. The evaporimeter is kept on a wooden frame of 100 mm thickness. The recorded gross evaporation rate is found to be 23% to 35% higher than the actual rate of evaporation from large bodies of water. To obtain the net evaporation rate, the observed gross values, E(mm/day) are multiplied by an average factor of 0.7 (F_1).

Fig 2.1: Standard Evaporimeter

The brine evaporation rate is lower than that of fresh water. It is proportional to the lower vapour pressure of brine. Another factor, F_2, is used to obtain brine evaporation rate from fresh water evaporation rate. The values of F_2 are shown in following table: (Table 1) Density – Evaporation Relationship for 0° to 30° Bé density brine of marine – origin.

Table 2.1: Density – Evaporation Relationship

Density °Bé	Evaporation	Density °Bé	Evaporation	Density °Bé	Evaporation
0	1.000	10	0.880	21	0.748
1	0.988	11	0.868	22	0.736
2	0.976	12	0.856	23	0.724
3	0.964	13	0.844	24	0.712
3.5	0.958	14	0.832	25	0.700
4	0.952	15	0.820	26	0.670
5	0.940	16	0.808	27	0.640
6	0.928	17	0.796	28	0.610
7	0.916	18	0.784	29	0.580
8	0.904	19	0.772	30	0.550
9	0.892	20	0.760		

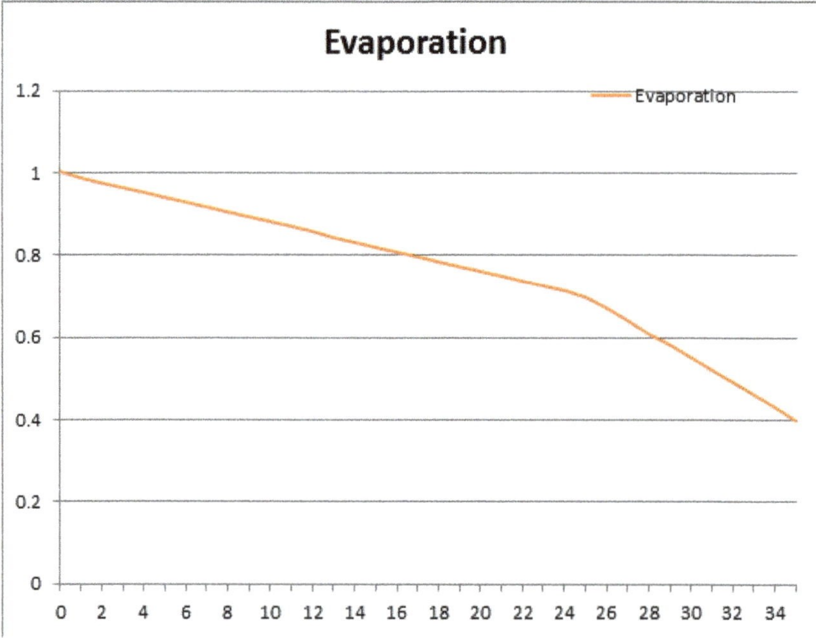

Fig 2.2: Density Evaporation Relationship

Basis: $F_2 = 1 - 0.012\,x$ (for 0° - 25° Bé)

F_2=1.450 – 0.03 x (for 25° - 30° Bé)

(Where x = density of brine in °Bé)

Brine evaporation rate = E x F_1 x F_2 mm (eq. 1)

Where

E = Gross evaporation rate of fresh water, mm

F_1 = Conversion factor

(Value ranging from 0.65 – 0.77, but taken as **0.7**)

F_2 = Conversion factor from above Table 1(Density – Evaporation Relationship) for the required density of brine.

There were found to be empirical relationships between evaporation rate (measured in evaporimeters, pans, ponds, etc.) and the meteorological- parameters wind and humidity. They were shown to be dependent on the difference between-the pressure of water vapour at the water surface and in the air above, together with an atmospheric turbulence factor closely related to wind speed affecting the air layer through which the water vapour was passing.

The brine of the crystallizing ponds is coloured a natural red by certain microscopic flagellate organisms, and also by bacteria, capable of thriving in it. These will help to absorb some of the radiation and so reduce the losses, but as they are in the form of a suspension rather than a true solution they cause a secondary loss by back-scattering. Some saltfield operators have considered encouraging the growth of the organisms by feeding them with nutrients, and so reducing the evaporation losses. Some actually add an organic dye to the brine with the object of entrapping all the incoming radiation. The dye most commonly used is of the 2-Naphthol Green type. Green is not the ideal spectral colour for optimum radiation absorption, but this particular dye has adequate absorptive power, and has a combination of properties leading to its choice for the duty. It remains dispersed in true solution in natural brine and it is reasonably resistant to fading in sunlight. Concerning the last, while it does slowly fade under constant exposure to the sun and atmosphere as do all organic dyes, it has better fading resistance than other dyes.

Knowledge of evaporation rate can be used in the design and operation of a solar saltfield. The simplest arrangement for the control of pumping or flow operations is to use a small, meteorological-pattern evaporation pan (preferably employing fresh water), its readings being equated to the evaporation of the various ponds by means of a pan "coefficient" and a salinity factor. Although involving an oversimplification, this method has been found quite satisfactory in practice.

Volume – Density Relationship

In a system where no solids separate out, the product of (d – 1) and y is constant where d is the density of brine in g/cm^3 and y is the volume. During the crystallization state, the volume first changes rapidly and then more gradually as the vapour pressure rises. Based on the two facts the density – volume relationship has been computed and is present in the following table (Table 2: Density – Volume Relationship).

Table 2.2: Density - Volume Relationship

Density °Bé	Volume	Density °Bé	Volume	Density °Bé	Volume	Density °Bé	Volume
1	3.5618	8	0.4236	16	0.1995	24	0.1246
2	1.7686	9	0.3738	17	0.1863	25	0.1188
3	1.1708	10	0.3340	18	0.1746	26	0.0721
3.5	1.0000	11	0.3013	19	0.1641	27	0.0521
4	0.8719	12	0.2742	20	0.1546	28	0.0403
5	0.6926	13	0.2512	21	0.1461	29	0.0330
6	0.5731	14	0.2315	22	0.1383	29.5	0.0280
7	0.4877	15	0.2144	23	0.1312		

Data from above both tables (Table 1 & Table 2) are used in calculating the area of individual compartment and volume requirement of desired density brine.

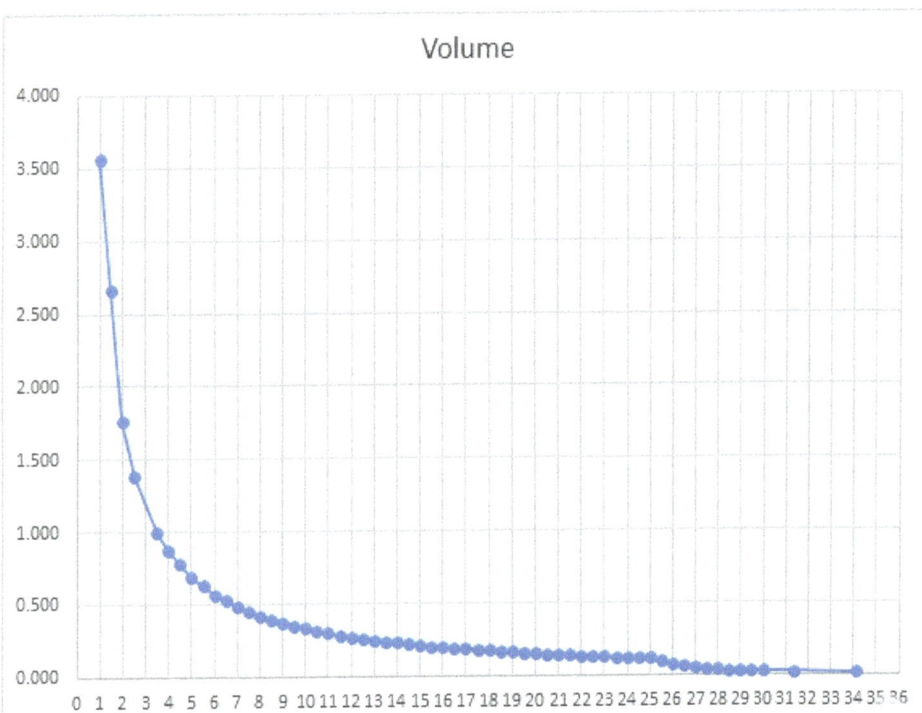

Percolation

Brine leakage from solar evaporation ponds constitutes one of the major factors in the design, operation and economics of salt recovery. Study of the nature of the soil used in lining the ponds is therefore an essential step before designing a salt works. The majority of the soil near the sea cost are alluvial in nature showing strong stratification. This property makes exact measurement of percolation very difficult. But a designer, with some broad guidelines, should be able to design a fairly reasonable plan.

Soil Classification:

Classification of soil based on the International Soda Pipette Method (Piper, 1950) is the most suitable guideline to predict the extent of leakage of brine. It is based on particle size classification in four types, depending on particle diameter. The mineral matter of the soil is separated from organic matter, gypsum, calcium carbonate and soluble salts. It is then classified into four groups of coarse sand by wet sieving and then into fine sand, silt and clay by the sedimentation method.

Coarse sand	0.2 – 2.0 mm
Fine sand	0.02 – 0.2mm
Silt	0.002 – 0.02 mm
Clay	Below 0.002mm

A gradation curve is obtained by plotting particle diameter on a logarithmic scale and cumulative percentage on an arithmetic scale. Porosity factor, n at field density is estimated by standard procedure. Average particle diameter, D_{50}, is read from gradation curve. Uniformity coefficient, C_u which is the ratio of D_{60} to D_{10} is also obtained from the gradation curve. Based on Kozney's equation (Kozeny, 1927) and it further modification (Joshi, 1970) the permeability coefficient is calculated with the help of the following equation:

$$k = \frac{n^3 (D_{50})^2}{(1-n)^2 (C_u)^{1.766}} \times 6.25 \times 10^3 \qquad \text{(Eq.2)}$$

Where:

k = permeability coefficient, mm/day

n = porosity factor

D_{50} = Average particle diameter at 50% on the gradation curve, mm

C_u = Uniformity coefficient, D_{60} / D_{10}.

Based on the particle size distribution in the three groups (sand, silt and clay) soil have been classified into three major classes and ten different types as shown in table below (Table 3).

Table 2.3: Classification of Soil and Their Percolation rate.

Class	Type	Percentage			Percolation of Saturated brine mm / day
		Sand	Silt	Clay	
Sandy	Sand	92–100	0–8	0–8	480.00
	Sandy loam	70–92	0–12	0–12	97.44
	Loamy Sand	62–65	5–25	8–35	5.78
	Sandy Clay	50–75	0–7	25–50	0.45
Loamy	Loam	50–75	10–25	10–25	0.36
	Silty Loam	0–75	25–100	0–25	0.24
	Silty Clay loam	0–50	25–75	25–40	0.20
	Clay Loam	35–70	8–25	25–40	0.15
Clayey	Silty Clay	0–35	25–60	40–75	0.12
	Clay	0–64	0–25	30–100	0.10

Percolation Rate

Using equation of the Permeability coefficient (Equation 2) the average rate of percolation of brine of 25°Bé (density of brine at the salting pond) is calculated and shown in the last column of the table (Table 3, The Table for Classification of Soil and Their Percolation rate). The first three soils have very high percolation rate. They are unsuitable for salt works. Next three are moderately suitable, where the last four are most suitable as far as loss due to percolation is concerned. The values of percolation are calculated from the average of each type of particles with ideal particles size distribution and smooth gradation curve. Highest value of the uniformity coefficient are assumed. It would be necessary therefore, to compute the percolation losses from actual analysis of the soil using Equation of permeability coefficient (Equation 2).

As the brine proceeds progressively from reservoir to crystallizer its depth decreases (Unless new brine is added) and the density and viscosity increase. Because these three properties have pronounced effect on the quantity of brine lost through the soil bed, appropriate corrections are made to arrive at absolute values of percolation.

Area Requirement:

The area of a solar salt works is subdivided into different compartments to facilitate the flow of brine and to control the density. The different evaporating ponds, their density range and average depth of brine are shown in the Table of Pond Details as follows (Table 4).

Table 2.4: Details of Ponds

Compartments	Code	Density Range °Bé	Average Density °Bé	Depth Mm
Crystallizer	A	25 – 29.5	27.5	150
Condenser I	B	23–25	24.0	150
Condenser II	C	17–23	20.0	200
Condenser III	D	14–17	15.5	250
Condenser IV	E	10–14	12.0	275
Condenser V	F	6–10	8.0	300
Reservoir I	G	3.5–6	4.75	400
Reservoir II	H	3.0–3.5	3.25	400
Reservoir III	I	2.5–3.0	2.75	400
Reservoir IV	J	2.0–2.5	2.25	400

The area of crystallizers forms the basis for computing the area of other evaporating ponds. The crystallizer area is directly proportional to the nominal production level and inversely proportional to the rate of evaporation and number of crystallization days. Once the crystallizer area is known, the relative area of other ponds can be computed.

Area Calculation:

To derive an empirical formula to calculate the crystallizer area, the following data of the Salt works is to considered, A model salt works with a production capacity of 5000 tons per annum is assumed.

Average daily evaporation (Oct to May) =7.4712 mm

Number of crystallization days = 150

Targeted production per year = 5000 tons

Daily average production = 33.333 tons.

Saturated Brine Requirement:

Tray experiments indicates that one liter of saturated brine (25° Bé) yields 222g of sodium chloride on reaching 29.5 °Bé or 1 ton of salt production will require 4.5 m^3 of saturated brine (on field practically requires 6 m^3). Assuming 90% efficiency of salt harvesting, the saturated brine required to produce 33.333 mt / day works out 166.67 m^3.

Crystallizer Area:

Volume of 25° Bé brine = 166.67 m^3

$$\textit{Volume left at } 29.5^\circ \textit{ Be} = \frac{166.67 \times 0.028}{0.1188} \text{ (Refer Table 2 : Density Volume reduction)}$$

$$= 39.28 \text{ m}^3$$

Volume of water to be evaporated = 166.67 – 39.28

= 127.39 m^3

Average evaporation rate of 27.25° Bé = 7.4712 × 0.7 × 0.6325

(Equation 1: Brine evaporation rate)

(Value of F_2 from Table 1: Density-Evaporation relationship)

= 3.308 mm/day

= 0.003308 m/day

Area = 127.39/0.003308

= 38510 m^2

Area of Other pond

Using the method described above, the area of evaporating pond compartments B to J is calculated and presented in following Table 5 : Area of Crystallizers and Other Evaporating Ponds

Table 2.5: Area of Crystallizers and Other Evaporating Ponds

Compartment	Code	Density Range ° Bé	Area m^2	Ratio of compartment area to crystallizer acre R
Crystallizer	A	25 – 29.5	38,510	1.0000
Condenser I	B	23–25	4,665	0.1211
Condenser II	C	17–23	19,445	0.5049
Condenser III	D	14–17	14,892	0.3867
Condenser IV	E	10–14	32,112	0.8338
Condenser V	F	6–10	70,930	1.8418
Reservoir I	G	3.5–6	1,21,395	3.1522
Reservoir II	H	3.0–3.5	47,664	1.2377
Reservoir III	I	2.5–3.0	66,303	1.7217
Reservoir IV	J	2.0–2.5	98,858	2.5670

Empirical Formula for Crystallizer Area

Using the data and assumptions made in the preceding paragraphs, an empirical formula is worked out to compute crystallizer area. All the factors are listed below:

1. Nominal annual production = T tones
2. Volume of saturated brine required per tonne = Q m^3
3. Volume of bittern left at 29.5°Bé per tonne = S m^3
4. Number of crystallization days per annum = D
5. Gross evaporation rate of fresh water/day = E mm

6. Conversion factor for fresh water evaporation = F_1 (Equation 1: Equation of Brine Evaporation rate)
7. Conversion factor for brine evaporation = F_2
8. Efficiency factor of salt harvesting = 0.9

$$\text{Area of Crystallizer A} = \frac{T}{D}\frac{(Q-S)}{0.9} \times \frac{1000}{E \times F_1 \times F_2} m^2$$

Substituting values of the known factor, i.e,

$Q = 4.5\ m^3$, $S= 1.0606\ m^3$, $F_1 = 0.7$ and $F_2 = 0.6325$.

$$\text{Area of Crystallizer A} = \frac{T}{D}\frac{(4.5-1.0606)}{0.9} \times \frac{1000}{E \times 0.7 \times 0.6325} m^2$$

Area of other Pond Compartment

Once the area of crystallizer is known, the area of the other pond compartments (B to J) is obtained by multiplying A with the values of R shown in the last column of Table 5.

$$\text{Area of any compartment (B to J)} = \frac{0.8631\ T\ R}{D\ E} \text{ ha} \qquad \text{(Eq. 4)}$$

Area requirement with percolation

Because brine is lost continuously due to percolation, makeup brine and additional area for it must be provided to achieve the targeted production. To compute the area of different compartments the following four assumptions are made based on practical experience in the field:

Percolation through the crystallizer beds is assumed to be zero, as the initial salt layer makes the bed almost impervious. However, in the beginning of every crop, some percolation is experienced which may amount to a fraction of a millimeter for a few days. If the salt works is laid on previous soils, the crystallizer area is given clay treatment, as one cannot afford to lose the saturated brine.

In the compartments B and C, density – viscosity correction not applied because the brine of the penultimate compartment (B) is very near to saturation. Compartment C is lined by the deposition of gypsum (17 to 23 °Bé′), which checks percolation to some extent.

In condensers D to E, an increase in depth and decrease in viscosity enhance the percolation 1.67 times compared to saturated brine of 25° Be′ with 150 mm depth. Similarly, in compartment F to J the ponds containing the least concentrated waters, the enhancement is 2.088 times. These two correction have been considered in calculating the necessary pond area.

Percolation rate of 0.5 to 3.0 mm/day with an increment of 0.5 mm are considered for calculation with 150 mm depth of saturated brine.

Area of Condenser B (23 - 25° Be')

To feed the required quantity of saturated brine to the crystallizers, an additional quantity of 23° Be′ brine equivalent to percolated (lost) brine has to be made up in order to maintain production. Obviously an additional concentration area also would be needed. Table 6: Area for B (23 - 25° Bé) for Different Percolation Rate, shows the volume of average 24° Be′ brine lost and total area required for each percolation rate. Data from Table 1 and Table 2 are used in these calculations.

Table 2.6: Area for B (23-25° Be') for different percolation rate

Percolation Mm / day	Volume of 23°Bé in m^3	Percolated volume of 24° Bé in m^3	Percolated volume of 23° Bé in m^3	Total volume of 23° Bé m^3	Area M^2
0.0	184.05	Nil	Nil	184.050	4665
0.5	184.05	2.332	2.453	186.503	4727
1.0	184.05	4.664	4.906	188.956	4789
1.5	184.05	6.966	7.359	191.409	4851
2.0	184.05	9.328	9.812	193.862	4914
2.5	184.05	11.660	12.265	196.315	4976
3.0	184.05	13.992	14.718	198.768	5038

Area of Compartment C to J

In a similar way, the brine requirement for the remaining compartments has been calculated from the preceding compartment area and is presented in Table 7 (Area of Compartment (m^2) for Different Percolation Rate

Table 2.7: Area of Compartments (m^2) for Different Percolation Rates

Compartment	Density °Be	0.0	0.5	1.0	1.5	2.0	2.5	3.0
A	25 – 29.5	38,510	38,510	38,510	38,510	38,510	38,510	38,510
B	23–25	4,665	4,727	4,789	4,851	4,914	4,976	5,038
C	17–23	19,445	20,575	21,704	22,836	23,966	25,099	26,229
D	14–17	14,892	16,394	17,899	19,403	20,907	22,411	23,915
E	10–14	32,112	37,590	43,171	48,553	54,035	59,513	64,995
F	6–10	70,930	91,897	1,13,651	1,33,497	1,54,712	1,75,658	1,96,609
G	3.5–6	1,21,395	1,72,114	2,24,265	2,73,260	3,24,282	3,75,013	4,25,732
H	3.0–3.5	47,664	69,141	91,205	1,11,986	1,33,611	1,55,094	1,76,576
I	2.5–3.0	66,303	99,235	1,33,004	1,64,968	1,98,200	2,31,033	2,63,9[illegible]
J	2.0–2.5	98,858	1,52,904	2,08,195	2,60,777	3,51,149	3,69,155	4,23,[illegible]

Effect of Initial Density of Seawater

If the initial density of seawater is below 3.5° Bé, i.e., the seawater is somewhat diluted by meteoric inflow, additional area must be provided to attain seawater concentration (3.5° Be'). The ratio of the area of the other necessary evaporation ponds to that of crystallizer will increase with lowering of the initial density. Considering this fact, such ratios have been calculated for four initial densities (3.5, 3.0, 2.5 and 2.0° Be') and for all the percolation rates (0.5, 1.0, 1.5, 2.0, 2.5 and 3.0 mm /day), as shown in following table 8.

Table 2.8: Area of Condenser + Reservoir for Different Starting Densities and Percolation Rate and Ratio of Condenser + Reservoir to Crystallizing Area

Percolation Mm/day	Initial Density	Area m^2			
		3.5 ° Be'	3.0 ° Be'	2.5 ° Be'	2.0 ° Be'
0.0	Area	2,63,439	3,11,103	3, 77,406	4,76,264
	Ratio	6.84	8.08	9.80	12.37
0.5	Area	3,43,297	4,12,438	5,11,673	6,64,577
	Ratio	8.91	10.71	13.29	18.56
1.0	Area	4,25,479	5,16,864	6,49,688	8,53,883
	Ratio	11.05	13.42	16.87	22.28
1.5	Area	5,02,400	6,14,386	7,79,354	10,40,131
	Ratio	13.05	15.95	20.24	27.01
2.0	Area	5,82,816	7,16,472	9,14,527	12,65,676
	Ratio	15.13	18.60	23.75	32.87
2.5	Area	6,62,669	8,17,763	10,48,796	14,17,915
	Ratio	17.21	21.23	27.23	36.82
3.0	Area	7,42,518	9,19,094	11,83,059	16,06,270
	Ratio	19.28	23.87	30.72	41.71

The extent to which the percolation rate and the initial density of seawater affect the area of reservoir and condenser is shown in Table 8. The ratio of area of reservoir + Condenser to area of crystallizer increase from 6.84 for 3.5° Be' initial density without percolation to 41.71 for 2° Be initial density with 3.0 mm / day percolation.

Area of Other Ponds with Percolation

To calculate the [illegible] condenser or reservoir for a given percolation rate, [illegible] the following formula:

$$A = \frac{0.8631\,T\,R\,R'}{D\,E} \quad \text{(Eq. 5)}$$

[illegible] nt in hacteares

T = Nominal annual production in tones

R = Ratio of evaporating ponds to crystallizers area (Table 5)

R′ = Ratio of area obtained from Table 7, $\frac{\text{Area for p mm percolation}}{\text{Area for zero percolation}}$

D = Number of crystallization days per annum.

E = Gross fresh water evaporation rate, mm/day.

An example given below shows the actual calculations.

Compartment under consideration – Condenser of 10 - 14° Bé

Nominal annual production – 1,00,000 mt.

Gross evaporation of fresh water – 6 mm/day

Number of crystallization days – 200

Expected percolation rate – 1.5 mm/day

$$A = \frac{0.8631 \times 1,00,000 \times 0.8338}{200 \times 6} \times \frac{48533}{32112} = 90.63 \text{ ha}$$

$$\text{Value of } R' = \frac{48533}{32112} \quad \text{(From Table 7)}$$

In the way, area of any other necessary compartment sizes can be worked out.

Sea Water Requirement

Daily requirement of seawater is based on three parameters, (1) Volume lost due to percolation, (2) Volume lost through evaporation, and (3) Volume charged in the crystallizers. The first parameter is almost constant, depending on the soil nature, but evaporation rate in winter is about 40% of the rate in summer. Therefore, in order to find out the daily requirement of seawater, the following formula is used:

$$V = Vp + Va \times (Ed / E) \qquad \text{(Eq. 6)}$$

Where:

V = Volume of sea water required on a particular day.

Vp = Volume percolated equivalent to sea water.

Va = Average daily requirement

E = Gross average Fresh water evaporation rate

Ed = Gross fresh water evaporation on the particular day.

In designing a solar salt works, evaporation rate of fresh water, number of clear days in a year, soil nature initial seawater density, and nominal production level and percolation rate are considered as governing factor.

Based on this data the area of crystallizers, condensers and reservoirs is calculated by employing empirical formulae. Volume density relationship and brine evaporation rate with respect to fresh water evaporation rate from the basis for computing pond area requirement and daily seawater requirement for an optimal lay-out of simple solar salt works.

In brine circuit designing that may some points at which we have to mix brine from two or more different brine circuits. At that location to predict the density of mixed brine can derived with using following formula.

Let us assume the output of brine circuit X is XVol and the density is XBe and the output of circuit Y is Yvol and density is YBe. Then the output density ZBe can predict with following equation.

ZBe = ((XVol x XBe) + (YVol x YBe))/ (XVol+YVol)

This is probable composition while all the conditions are ideal as per our calculation factors used in our calculation like evaporation, rain fall, pumping quantity, no wind effect, etc.

Fig 2.3: A well design Salt Works. Tata Chemicals Limited Mithapur, Gujarat. (INDIA) Capacity – 3.2 Million Metric Tons per annum.

3

Civil Aspects in Construction of Solar Salt works

The important civil engineering aspects that should be considered as part of the feasibility study for developing Solar salt works are as follows.

1. Soil Characteristics.
2. Natural Gradient and topography of site.
3. Storm water drain
4. Availability of construction materials.
5. Approach Roads.
6. Availability of Construction Power and Water.

1. Soil Characteristics.

A detailed geotechnical investigation of site is required for deciding the soil characteristics.

Soil bearing capacity: This will decide the depth to which the foundations of the Pumping stations, Buildings and the embankments need to be taken. When soil strata with required strength is not available at shallow depths the foundations have to be taken deeper. This can have a major impact on the Project cost.

Soil Characteristics for embankment material: It is required to test the samples of the soil which is planned to the used for the construction of embankments. The properties like Shear strength. Optimum moisture content, maximum dry density, Cohesion and angle of internal friction are ascertained by testing he samples in the Geotechnical Laboratory.

2. Natural Gradient and topography of the site

A detailed site survey needs to be carried out for deciding the contours. Natural gradient if favorable to the flow can save cost and vice a versa. The flow of water should be planned to make use of the natural gradient.

3. Storm water drainage

From the site survey and the reconnaissance survey the natural flow pattern of the storm water should be understood. The salt works are located at the sea cost and at lower levels hence there are maximum chances of obstruction of the storm water passage. If the topography is favorable then a strip of land is left unutilized on the periphery which acts as a storm water drain else separate storm water channels have to be provided on the periphery of the salt works. Proper estimate of the catchment area and the rainfall intensity should be done for deciding the cross section and the gradient of the storm water drain.

4. Availability of construction materials

Soil from the bed of the ponds is normally used for the construction of the embankments of the salt works. Thus the material for embankment construction is most of the times available within the site. In some cases when the material from the bed is not adequate for embankment construction it will have to be carted from far of locations which will add to the cost of construction. The other important construction material is stones for the road work as well as the pitching of Embankment side slopes. Availability of god quality stones from close location will have to be explored.

5. Approach Roads

If the salt works located away from a highway or a good motorable road the approach road up to the site will have to be constructed for construction activities as well as for the transport of salt.

6. Availability of Construction Power and water

This is an important requirement of any construction site and should be accounted for and arranged in time.

The crystallizers, condensers and reservoirs of solar salt works are designed on the basis of empirical formula. The governing fctors in design are Evaporation rate, Number of clear days in a year, Initial sea water density, Percolation rate, Rain fall. Out of the one characteristic that can be controlled to a certain extent by proper engineering is percolation loss.

Percolation loss from pond bed

Percolation is complex phenomenon and proper estimation of the percolation loss is important to get the expected yield from the saltworks. The percolation loss mainly depend on the Voids ratio of the soil bed and the density of the fluid. The voids ratio is dependent on the size of the soil particles. Based on the particle size the soil is classified as:

Coarse Sand	0.2 - 2.0 mm
Fine Sand	0.02 - 0.2 mm
Silt	0.002 - 0.02 mm
Clay	Below 0.002 mm

Depending on the particle size the percolation losses range from 480mm/day in coarse sands to 0.1mm/day in clay. The natural soil has a mix of sand, silt and clay. There is also soil stratification and hence computation of actual soil characteristic is important for design of ponds.

Methods to reduce Soil percolation

Soil percolation can be reduced by Soil compaction, Soil Replacement, Lining of pond bed, Soil/LDPE barrier on the periphery.

Soil compaction

Compaction of the soil bed using Vibro rollers or static rollers will reduce the voids ratio of the soil. This will result in reduction of percolation. This is simple and low cost option where the soil is fairly good.

Soil Replacement

When the soil is having porous bed like coarse send, it will have to be replaced with clayey soil or other available soil which has better particle size distribution. A layer of minimum 300mm need to replace for good result. If the levels are favorable to save costs an additional layer of good quality soil is laid on the existing bed.

Lining of Pond bed

Use of LDPE liners is an effective method to avoid percolation through soil. The success of this method depends on avoiding damage to the LDPE sheet during and after construction. Should take due care while laying of sheets with proper anchor to avoid floating while feeding of liquid. If possible compact base floor of pond before the laying.

Fig 3.1: LDPE lining at Pond Bed

Soil/LDPE barrier on the periphery:

This method is useful when there is a good stratum at reasonable depth of about 2 meters. A trench is dug on the periphery and is filled up with good quality clayey soil well compacted in layers. Fixing of LDPE sheets on the vertical face of trench before back filling will give excellent result.

Fig 3.2: Soil/LDPE barrier on the periphery:

Civil Construction

The civil construction activity for development of solar salt works mainly consists of Site clearance & grading, Embankments, Roads, Channels, Pumping Stations, Brine Pipelines, Culverts, Buildings like Office, weigh bridge, Operator cabins, Rest Shade, Toilet Block etc.

The estimation of cost of each of above will share as follows.

Sr. No.	Description	% Cost
1	Site Clearance and Grading	2.70
2	Embankments	43.13
3	Stone pitching of embankment slopes	40.43
4	Intake Channel	1.08
5	Roads	6.06
6	Pumping stations	1.75
7	Culverts	1.89
8	Buildings	1.07
9	Storm water drain	1.89
	Total	100

Earthen Embankments

To divide the area into separate compartments and to retain the sea water/Brine up to required depth and to make approach roads we need to construct earthen embankments.

As we have seen above the higher cost sharing by earth embankments and stone pitching of embankment slopes.

To construct proper embankment we need to know soil characteristics below embankments by conducting soil investigation at laboratory mainly for soil bearing capacity, and the material of construction of the embankments to be tested in the laboratory for its characteristics mainly Maximum density and optimum moisture content.

Embankment Design Concepts:

The design of earthen embankments is based on the principle of geotechnical engineering. Since the embankments in salt works are of lower heights and do not retain water up to higher depth, they do not call for detailed slope stability analysis and design. The Size of Embankment in salt works is decided based on (1) The top width required for the movement of vehicle. (2) The side slopes are generally kept at a slope of 1V:1.5H. (3) Depth of the embankment to be taken below the Natural ground level based on the strength of the soil at various depths below the ground. The founding strata should have enough bearing capacity to support the dead and live loads on the embankment.

Embankment's Design

Fig 3.3: Lining of embankment with Fly ash

Embankment's Construction:

Precautions to be taken while construction of Embankment to avoid damage due to rains or due to traffic movement during use are as follows.

- Compaction: In any type of embankment construction the most important factor is proper compaction. The construction should be done in layers of 1 to 2 feet. Each layer should be watered and compacted with vibro-rollers. Only enough amount of water should be added. The proctor test for soil compaction should be carried out at random after compaction of each layer.

- Side Slope: A mild slope reduces the weight of the earth tending to slide and hence improves the slope stability. The side slope of 1V:1.5H.
- Protecting the Toe: A berm at the toe of the embankment increase the resistance to sliding and avoids base failure.
- Proper Drainage: The embankments on the periphery or the once that retain water only on one side should have a proper seepage channel. This will reduce the seepage force and also help in dissipating the pore water pressure.
- Soil Stabilization: Soil on the slopes can be stabilized by adding cement or other compounds. Use of Geo Synthetics for Slope stabilization is a reliable technique but is costly and is not feasible for Salt-works.
- Lining of Slide Slopes: The lining of the side slopes is an effective way of protecting the side slopes. Unlined embankment slopes get damaged during rains and call for regular maintenance. All the important embankments should be lined for better life and to avoid contamination.
- Lining with Stone Pitching: When good quality stones are easily available from short distance this is the most cost effective option. The thickness of the stone pitching is normally kept as 230mm. The stones are hand place by skilled masons. The stones should be good quality and should not have any effect when they come in contact with brine. A small toe wall at the base is must for stability of the pitching and for protection of base.

Fig 3.4: Stone Pitching

- Lining with Fly ash:

  Lining with Fly ash is a good solution and will also as sink for fly ash generated in Thermal Power Plants. The bottom as which is general not useful in cement plants is the most suitable material. The fly ash is transported to site and is arranged in circular pits. The pit is flooded with water for 24 hours. The surface of the embankment to be lined is properly dressed and compacted manually and proper line and level is maintained. Fly ash mortar is prepared by simply adding required quantity of water. The mortar is applied on the prepared surface of the slope in the same manner as the building plaster. The thickness varies from 50mm to 75mm. It is then cured for 7 days. Adding about 1 to 2% of cement will improve the quality and avoid cracking. The total cost of lining will be about Rs. 50/sq.mt. (1 USD)

Fig 3.5: Lining of embankment with Fly ash

Lining with Salt:

In crystallizers lining with salt is the best option. In crystallizer we can spread off grade salt on the surface of crystallizer's embankments. Salt will protect embankments by washing and will maintain density of brine while in rain. At time of rain the salt at slope of embankment dissolves and enriches density of brine. This will help to maintain salt bed thickness in monsoon season. At the time of harvesting 5 to 6 meter of salt parallel to embankment should harvest and spread to embankment to maintain thickness which washed off at the time of rain every year.

Fig 3.6: Lining with salt.

4

Recovery of Dissolved Substance

Sea Water

From the different analysis given in the following tables it will be noticed that general nature of the constituents in the different sea waters throughout the world is very nearly alike. The difference is in degree but not in the character of the constituents. The quantities of the principal salts vary within very narrow limits, and the total of solids dissolved in most of them is also within very narrow limits.

Composition of dissolved salts in Sea Water

Grams of Salts per liter of brine as a function of brine density g/liter of brine at 22.2° C

Sample 1:

Bé	3.5		
Sp.Gr.	1.0247		
$CaSO_4$	1.419	KCL	0.711
$MgSO_4$	2.138	NaBr	0.085
$MgCl_2$	3.384	Total Salts	36.04
NaCl	27.27	H_2O	989.66

Sample 2:

Bé	3.5	°Bé
Sp.Gr.	1.0247	
NaCl	27.84	Gpl
$MgCl_2$	3.528	Gpl
$MgSO_4$	2.32	Gpl
$CaSO_4$	1.31	Gpl
KCl	0.675	Gpl
NaOH	0.000	Gpl
Na_2CO_3	0.000	Gpl
Na_2SO_4	0.000	Gpl
Bromide	0.066	Gpl
Phosphorus	0.07	Mg/l
Zinc	0.01	Mg/l
Copper	0.003	Mg/l
Strontium	0.008	Mg/l
Boron	4.8	Mg/l
Silicon	3	Mg/l
Fluoride	1.3	Mg/l
Rubidium	0.12	Mg/l
Lithium	0.2	Mg/l
Arsenic	0.003	Mg/l
Tin	0.003	Mg/l
Lead	0.003	Mg/l
Aluminium	0.01	Mg/l
Barium	0.03	Mg/l

Recovery of Dissolved Substance

Substance Dissolved in Sea Water and Brine:

From the analysis of sea water given above it will be found that the principal constituents of the total dissolved solids in sea water are following:

Calcium Carbonate.

Calcium Sulphate.

Sodium Chloride.

Potassium Chloride.

Magnesium Sulphate

Sodium Sulphate.

Magnesium Chloride.

Magnesium Bromide.

Besides these salts there is a large number of others occurring in very small quantities in sea water, e.g., Iodides, Fluorides, Phosphates, Nitrates, Iron, Silver, Gold, Copper, Lead, Arsenic, Zinc, Nickel, Lithium, Rubidium and Cesium; but as these substance occur in infinitely minute quantities, they are of no technical importance in the recovery of substances from sea water.

The dissolved substances give to the sea water their characteristic saline taste. There is limit to the solubility of each of these substances. This limit depends upon the temperature of the liquid and presence of other dissolved salts. In each individual case, water at a given temperature will dissolve only a certain quantity of the substance. When this quantity is dissolved, it will dissolve no more of that particular substance. The solution in that case gets saturated. In order to recover the dissolved substance, water which is not chemically changed by the dissolved substance must be removed by evaporation either spontaneously in the open air or by the application of heat. Most of the solids are dissolved in larger proportions by increase of temperature, but in the case of Sodium Chloride, its solubility remains practically constant at higher temperature.

If the saturated solution is lowered in temperature or its volume reduced by evaporation, the equilibrium is disturbed, and the solution in that case holds more salt than it is able to retain in its normal solubility conditions. The solution is at this point super-saturated, and the salt in excess will be thrown out of the solution as a solid until again a stable solution is formed under the altered conditions. This process becomes complicated if the solution contains more than one dissolved salt. The separation of the dissolved solids as the evaporation proceeds is in a fixed order. The least soluble salt separates first and it is followed by the other salt in the order of its solubility. By regulating the rate of evaporation it is thus possible to separate out each salt as it reaches its point of saturation. This process known as fractional crystallization is applied in the separation of the various salts dissolved in sea water. In certain cases two salts separate out simultaneously. In such cases the mixture of salts is re-dissolved in fresh water, and the solution again concentrated and the salts re-crystallized. By this process of re-crystallization, pure salts are isolated. If, when the solution is super-saturated a minute crystal of the salt is introduced into the solution by way of seeding or inoculation, the process of crystallization starts spontaneously in the solution. In many cases if the super-saturation is carried far enough, the crystallization will grow without the necessity of inoculation. When inoculation is necessary, the solution is in the metastable state, and when no inoculation is required and where the crystallization is spontaneous, the solution is in the labile state. In the concentration of sea water we have these three states occurring between the following densities:

From 3.5-17° Bé the solution is unsaturated.

From 17-25° Bé it is saturated as far as Calcium Sulphate is concerned.

From 24-25° Bé the sea water or brine is in the metastable state.

From 25-29.5° Bé it is in the labile state as far as Sodium Chloride is concerned.

Evaporation:-

The change in water from its liquid to the gaseous state is brought about in increasing the molecular activity by the application of heat. The molecules of water are in a state of perpetual motion.

The energy of this motion depends on the temperature and pressure. It is continuous, but increase if heat is applied to the liquid. When the molecules in motion are in liquid they are attracted by other surrounding molecules, but when these molecules reach the surface, some of them are moving with sufficient velocity to escape from the surface of the liquid into the surrounding space and mingle with the air over the liquid. Some of the molecules that have thus escaped have not sufficient energy to remain in the space and are thrown back in the liquid. Over the liquid there is a certain layer of molecules in gaseous form in constant struggle to go out of and come inside the liquid; the process of coming out of the liquid is known as evaporation. In short, water appears in gaseous state in the from of vapor at its surface. As the molecules are projected from the surface into the air they have to encounter a certain resistance in their passage exerted by the pressure of the other molecules of water already existing there. If the pressure is great, a number of molecules strike back on the surface and return to the water till for a given temperature a maximum vapor pressure is obtained. The air above the liquid in that case is saturated. But if, as the molecules come out of the liquid they are rapidly carried away by a current of air, they do not get a chance of going back to the liquid state and their place is taken by fresh molecules coming out of the water. The cause therefore which influence the rapidity of evaporation of water in open surface are:

1. The temperature which increase the molecular activity;
2. The quantity of the same vapor in the surrounding atmosphere;
3. Renewal of this atmosphere, and
4. Extent of the surface of the evaporation inducing a large number of molecules to strike out of the water as vapor at the same.

If evaporation is to be accelerated, it is necessary that:

1. The molecular activity or temperature should be high.

   The Temperature of the brine would depend on the temperature of air.
2. The surface exposed should be as large as possible.

   If the brine is stationary the evaporation will be slow as the area exposed would be constant. If instead of this static condition, the brine is kept moving in thin layers, a very large surface would be exposed to the air and the evaporation in the latter case would be very rapid. The principal of a thin moving film is widely applied in chemical engineering for concentration of liquids. The exposed surface of liquid may be considerably increase by a slow trickling over a pile of pebbles or by spraying from great height. Both these methods are used in the rapid concentration of brine.

3. There should be wind.

   Wind is the greatest factor in the removal of saturated vapor from evaporating surface. Wind also increase area exposed by creating waves and ripples. By increasing the current of air and by increase the surface a maximum rate of evaporation may be obtained.

Solar Evaporation

In tropical and sub-tropical countries salt is recovered from sea water and brine by solar evaporation. If local conditions as regards supply and sea water of brine, soil level and transport are favorable, salt manufacture by solar evaporation is far more advantageous than by any other means. Solar evaporation requires the general fulfillment of all the above factors, which accelerate the evaporation of water in open surfaces. They are best studied from meteorological observations of the place at which it is proposed to locate the factory/salt works. The factors of importance are:

1. Temperature of the air.

   The temperature of the air depends upon the altitude of the sun, latitude, elevation, distance from the sea and character of the winds, and also on the amount of clouds and rainfall.

2. Pressure, direction and velocity of the wind.

   Distance from the sea and direction and character of the prevailing winds are of great importance in the location of salt works.

3. Moisture in the air.

   The chief factors determining the amount of moisture in the air are Distance from the sea and character of wind. When air takes up all the moisture which it can take up at a given temperature it is said to be saturated. With an increase in the temperature the saturated air will take more of the moisture. The moisture in the air is determined by the wet and dry bulb hygrometer. When air is cooled gradually, a point is reached, when dew begins to form. Relative humidity is the ratio of the pressure of the aqueous vapor actually present at a particular temperature to the pressure of the vapor which would be present if the vapor were saturated at the same temperature.

4. Extent of cloudy skies.

   The amount of cloudy skies is estimated in terms of the proportion of the sky covered with cloud by numbers ranging from 0-10. A sky free from cloud is denoted by 0 and the overcast sky by 10. The estimate gives by Meteorological Department.

5. Rainy days.

   The rainfall and rainy days are determined from the average of the records of the Metrological Department and self data collection. Rainy day is counted when at least 1/10″ rain has been measured.

Absorption

There are no regular observation taken of loss in brine by absorption in channels, condensers and pans. In salt works the heavier density brines are not so rapidly absorbed as in the case of fresh waters. The absorption is considerably checked by the separation and deposit of Calcium Sulphate. For this reason at the end of the season and in older works there is very little loss of brine by absorption. In well-constructed salt field ten percent is usually allowed for the loss by absorption.

Crystallization of Salt:

When solid separates out on cooling or concentration of a saturated solution it assumes regular geometrical forms bounded by plane face. These solid symmetrical forms are known as crystals. Every chemical substance of definite composition which crystallization from solution has a specific crystalline from characteristic of that substance. This form never changes. "In whatever manner, or under whatever circumstances a crystal may have been formed, whether in the laboratory of the chemist or in the workshop of Nature, in the bodies of animals or in the tissues of plants, up in the sky or in the the depths of the earth, whether so rapidly that we may literally see its growth or by the slow aggregation of its molecules during perhaps hundreds or perhaps thousands of years, we always find that the arrangement of the faces of the crystals, and therefore its other physical properties are subject to fixed and definite laws." When a crystal is suspended in a saturated solution by means of a fine thread it grows uniformly in all directions and the ideal form is obtained, but under ordinary conditions there are various factors which influence the growth of the crystals:

1. If the concentration of the solution is not uniform, the growth of the different planes is not uniformly rapid and the crystal develops abnormally.
2. If the crystal rests on one plane on the surface of the vessel, then that face is not fully developed. Flat crystals parallel with the bottom of the vessel result. The formation of flat blocks of salt is due to the crystals not being disturbed in the liquid during their growth by accretion. If the crystals and the solution be stirred from time to time with raker or any other as in the case of Baragara salt a uniform growth in all directions will result.
3. A cubical crystal sometimes grows rapidly in one direction resulting in needle-like or bar-like form. Sometime only two faces grow and in that case the growth is like flacks or plates.

4. The formation is rapid if the solution is very concentrated. The temperature and concentration conditions of solution must be determined for each salt to obtain the best result.
5. Rapid crystallization result in the formation of smaller crystals.
6. The growth may be changed by the presence of other dissolved substances.
7. Crystallization in motion gives better and more rapid results than crystallization in repose. The crystals thus obtained are purer than crystals obtained in static condition.

Common Salt Crystals

Sodium Chloride forms symmetrical cubic crystals having nine planes of symmetry. Salt crystals obtained by spontaneous evaporation of sea water or brine without much disturbance are conical or hopper shaped. These crystals grains can be altered to some extent by constant stirring which separates the tiny cubes forming the cone or hoper and also by the addition of alum and the presence of magnesium sulphate or calcium sulphate. The effect of these substance is to form very hard large coarse crystals. When the salt crystals are formed in the presence of soluble sulphate as in the case of salt from sea water or brines or when

Fig 4.1: Salt Crystals

alum is added, the salt has the first tendency to crystallizing in hollow cubes in the surface of the water. The cubes grow gradually by the addition or the other smaller crystals around them. The growth is in the shape of hopper and when the aggregate is sufficiently heavy, it settles down to the bottom of the pan. If the aggregate is disturbed by stirring with the help of a raker or a spade the crystals separate out, and each one in turn begins to grow.

Fig 4.2: Manual Raking activity

If, on the other hand, a fine grained salt is required gelatin or grease is added. Salt, if badly prepared has the property of becoming hygroscopic owing to the presence of slight impurities of magnesium chloride which remain coated on its surface after it is removed from the bitterns in the pan. If the salt crystals are well washed the hygroscopic property is considerably reduced. However carefully salt is prepared it shows a tendency to stick together in lumps. Under identical conditions of temperature, concentration and substances in solution salt will separate in only one form of crystals. The size and the general form will depend entirely on the mode of preparation. Given brine of the same composition, it is possible to manufacture any grade of salt by variation in the method of recovery.

Salt by manual harvesting.

5

Technology of Salt

The process of obtaining Salt (NaCl) and other substance which are dissolved in sea water or natural brines is entirely one of gradual evaporation and factional separation of solids at different degrees of concentration. Most of the natural brines contain the same salts as those found in sea water. If these brines are reduced in strength to the same density of sea water, the composition of the two is very nearly alike. The weight of the total dissolved solids are in both cases within very narrow limits. The graphical representation follows the relation between Volume and Density of the sea water at different of concentration.

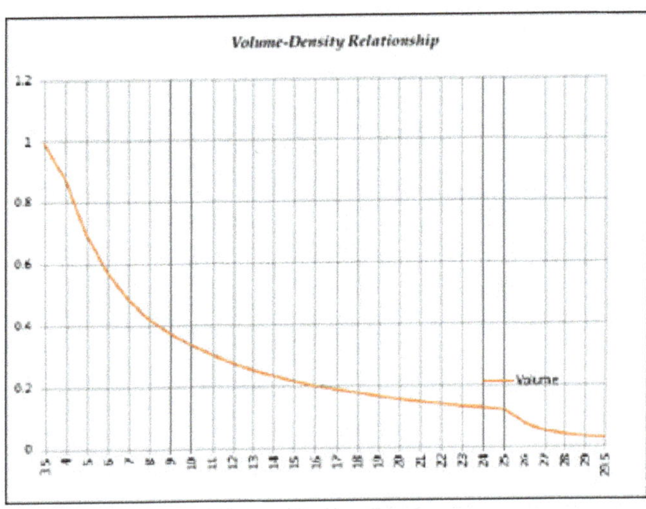

Density of Liquid

The density of a liquid is expressed in different terms. Specific gravity is the number, which expresses the relation of the weight of a given volume of the liquid to the weight of the same volume of distilled water at a temperature of 4° C. In order therefore to calculate the specific gravity of a body it is sufficient to determine its weights and that of an equal volume of water and then to divide the first weight by the second, the quotient is the specific gravity of the body. In the usual works practice, specific gravity is determined by means of a hydrometer or *SpGr* bottle. As specific gravity is not the same at different temperatures, the hydrometers are marked with degree Centigrade or Fahrenheit at which they are calibrated.

In salt works practice, Beaumés hydrometer is universally employed. The graduation of this instrument is made as follows. It is so constructed that when immersed in pure distilled water, the stem dips in the water nearly at the top. This point is marked zero. Each division indicate one percent dissolved salt. The graduation is continued to the stem up to 40 divisions in salt works practice.

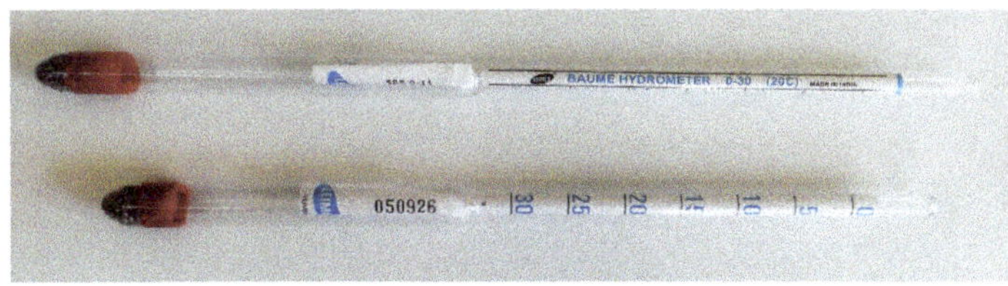

BeaumésHydrometer : 0°Bé – 30 °Bé

Fig 5.1: Beaumés Hydrometer : 0°Bé – 30 °Bé

There is another procedure to measure salt in brine. In some places Specific Gravity unit is using to measure salt in brine instead of Bé. For the conversion of the reading by these different hydrometers in to Specific Gravity the following formula used:

$$\text{Specific Gravity} = \frac{145}{145 - Bé}$$

When the solution contains other substances besides sodium chloride or salt, the hydrometer readings do not give a correct idea of the percentage of sodium chloride dissolved. But for all practical purpose the Beaumés hydrometer is of considerable value in salt works.

Concentration of sea water and brine Stages:

The process of concentration may be divided into seven stages. Each stage represents a distinct change in the resulting liquid.

1. Between 3.9° Bé – 10.0°Bé
2. Between 10.0°Bé – 17.0°Bé
3. Between 17.0°Bé – 24.5°Bé
4. Between 24.5°Bé – 29.5°Bé

5. Between 29.5°Bé – 35.0°Bé
6. Between 35.0°Bé – 37.0°Bé
7. Between 37.0°Bé – 38.5°Bé

First Stage:

Fig 5.2: First Stage: Painted Stork (Mycteria leucocephala)

Seawater is generally found to have a Specific gravity of 1.025 to 1.029 at 25°C corresponding to 3.5 to 4°Bé.

By gradual evaporation of sea water the original volume is reduced to 37 percent when the density reaches 10°Bé. Up to 10°Bé the first salt that is thrown out from the solution is Calcium Carbonate. It is extremely insoluble in water. It solubility at 30° C is only 0.0052 grms. In 100 grms. of water. As Calcium Carbonate is abundantly found in nature, its separation and recovery in salt works is of no practical value.

Second Stage:

Fig 5.3: Greater Flamingo : Phoenicopterus roseus
Generally observed in 2nd Stage

After the separation of Calcium Carbonate the liquid remains unsaturated till it reaches a density of 17°Bé. The original volume of the sea water is now reduced up to 20 % or one-fifth of the original. Gypsum and the remaining Calcium Carbonate and sometimes also Magnesium Carbonate are very often noticed to separate out at 12°Bé

Third Stage:

Fig 5.4: Artemia Salina : Observed in 3rd Stage

When the concentration has reached 17°Bé Calcium Sulphate begins to separate out as gypsum $CaSO_4 2H_2O$. The separated gypsum at first floats on the surface of the liquid as a thin gray film and when it has sufficiently accumulated, settles down to the bottom of the pan, carrying along with it a little salt which just begins to come out when the major portion of the gypsum cease to separate. The separation of gypsum continues up-to 25°Bé. The total amount of Calcium Sulphate originally present in the seawater is not entirely thrown out of the solution even when the brine reaches a concentration of 25°Bé as will be seen from the following table:

Separation of Calcium Sulphate

°Bé	Remaining dissolved	Total deposited	Percentage removed
3.5	1.749		
25.0	0.283	1.466	83.8
28.5	0.015	1.734	99.1
32.4	0.000	1.749	100

Source : Study and analysis by Manzella

Fig 5.5: Drained condenser for Gypsum harvesting.

Only about 84% of the dissolved Calcium Sulphate separate out during the second stage of evaporation. When natural brines are found having a density between 17° Bé - 25° Bé, they are generally deficient in Calcium Sulphate. In large sea salt factories Calcium Sulphate gradually accumulates in the condensers or preliminary evaporating basins. For every 100 tons of salt obtained from the seawater 4 to 5 tons of gypsum obtained. In the absence of any commercial demand or technical use, it can spread out in the making of roads and in making impervious floor or bottom of newly constructed crystallizing beds. While using at crystallizing ponds needs high care and high precision while harvesting the salt to prevent mixing with harvested salt, otherwise it will increase calcium impurities in harvested salt.

Fourth Stage:

This stage begins at 24.5°Bé. Pure salt solution is saturated at 26°Bé when the percentage of salt in solution is 26.5. But in the presence of other dissolved salts, the solution behaves as if it is saturated at 24.5°Bé, when salt begins to separate out. As the evaporation proceeds for every 100grm of water evaporated from the saturated solution, 36.5 grm of sodium chloride thrown out of the solution. The solubility of sodium chloride begin 36.5 grm at 30° C in 100 grm of water, the percentage of salt present in the saturated solution would be 26.74% by weight. Thus for every 100 grm of saturated solution we have 73.26 grm of water and 26.74 grm of salt.

Table 5.1: Composition of Dissolved Salts (in gram per 1000 ml) in Brine during progressive Evaporation.

Be'	Sp.Gr.	Dissolved Salts, in gms. per liter (gpl)						Total	
		$CaSO_4$	$MgSO_4$	$MgCl_2$	NaCl	KCl	NaBr	Salts	H_2O
3.5	1.0247	1.419	2.138	3.384	27.27	0.711	0.085	36.04	989.66
10.74	1.0800	4.499	7.157	10.710	89.02	2.414	0.278	114.10	965.90
15.54	1.1200	4.011	11.360	16.580	138.85	3.765	0.435	175.05	954.95
20.00	1.1600	2.898	15.690	22.740	190.72	5.159	0.596	237.79	922.21
25.17	1.2100	1.606	21.130	30.810	256.81	6.953	0.802	318.17	891.83
26.00	1.2185	1.400	22.060	32.280	268.38	7.266	0.840	332.29	886.21
28.06	1.2400	0.817	50.820	74.050	206.45	16.630	1.910	350.79	889.21
29.00	1.2500	0.595	63.240	92.710	180.34	20.760	2.390	360.15	889.85

From above table it will be noticed that the solution remains between 25 and 26° Bé for a very long time. This rise in density is not rapid on account of the separation of the dissolved salt. The density begins to be appreciably higher than 26 after the separation of over 50 % of the dissolved salt. The density fluctuates during the day and night. When during the day the evaporation is rapid under influence of the sun and the winds, the brine become slightly supersaturated; but when night falls and the temperature of the air is lower there is a drop in the temperature of the brine due to its cooling, and the consequent separation of the excess of salt from the super-saturated solution. This drop in the readings of the density taken in the evening of the previous day and the next day in the morning is very noticeable. As the concentration proceeds, more and more salt is thrown out of the solution till the brine reaches a density of 28.5° - 29°Bé. It is well known that the total quantity of salt originally present in the brine is not entirely separated when the density has reached 28.5° Bé. The salt obtained up to 28.5° Bé is generally of the following composition:

	Brine/Subsoil Salt	Sea Salt
NaCl	97.40	97.1
$CaSO_4$	0.58	1.10
$MgSO_4$	0.15	0.23
$MgCl_2$	0.08	0.04
Insoluble Residue		0.05
Water	1.71	1.40
	99.92	99.93

These typical analysis show that Sodium Chloride is associated with other salts, such as Calcium Sulphate, Magnesium Sulphate and Magnesium Chloride. Though Potassium Chloride is not shown in these two typical analysis, it is also found to the extent of 0.012%. Magnesium Chloride renders the salt hygroscopic and presence of the moisture over one percent is accountable to this salt. Magnesium Chloride also intensifies the salty taste. If the salt is washed with weak brine or fresh water, most of the above impurities, which are found in the residual containing of the bittern remaining on the surface of the crystals, are removed and the salt thus washed is 99% pure. The fourth stage is completed when the concentration reaches to 28-29°Bé. The concentration is not carried further, as the salt that separates out at this higher concentration is very impure and difficult to separate from the adhering bitterns. For this reason, at all the salt works, the process of recovering salt is almost stopped when the concentration has reached 29°Bé. At this stage the remaining bittern is only 2% of the original volume and its weight is partially the same as the weight of 90-94% salt recovered. The remaining liquid is known as bittern. Ballart, the French Chemist, was the first to suggest the recovery of these valuable by-product from bittern.

Fig 5.6: Crystallizer observed in Fourth Stage: Pink colored due to *halophilic* microorganisms

Fifth Stage:

When the bittern drained out from the crystallizers at 29.5° Bé that contain various salt in solution. These salts are at 29.5°Bé, very near their point of saturation. For this reason the bitterns require very careful treatment. As the original volume of sea water has contracted to 2% when it has reached 29.5°Bé. It will be more convenient to take the bittern of 29.5°Bé. as the starting liquid for the calculation of volume the subsequent stage of evaporation. As the deposits obtained up 36°Bé consists mainly of sodium chloride with a small percentage of magnesium and potassium chloride and sulphates. It will noticed from the solubility curves that magnesium sulphate and potassium chloride are soluble practically to the same extent at the ordinary temperature of the air (25°C). Their separation from saturated solution is very difficult.

Fig 5.7: Crystallizer of 5th Stage. Bittern of 35°Bé

Sixth Stage:

This is a very important stage in the recovery of by-products. At this stage potassium salts and magnesium sulphate and also sodium sulphate are recovered if the process of separation is worked out carefully. From the solubility curves and the table of solubility it will be noticed that magnesium sulphate and potassium chloride are of identical solubility. If the bittern of 35°Bé allowed to cool spontaneously during winter till it reach a temperature of 11-12°C when magnesium sulphate separates out. In order to prevent the formation of carnallite which is likely to deposit, the temperature is very carefully maintained at 12°C and not allowed to go below this point. The magnesium sulphate crystals are obtained. If after removal of the deposit of magnesium sulphate the bittern is further cooled to 6°C and under by means of a refrigerating plant, a double decomposition takes place between sodium chloride and the remaining magnesium sulphate. Sodium chloride must be in slight excess than theoretically required for the reaction. If it is deficient it must be made up by the addition of the required quantity. The double decomposition results in the formation of sodium sulphate ($Na_2SO_4 10H_2O$) and magnesium chloride; at the end of this stage the original bittern is reduced to 57 % of its volume.

Fig 5.8: $MgSO_4$ Crystals

Fig 5.9: Salt crystallized between 12° C to 6° C

Seventh Stage:

This stage is in every way as important as the last. If it is desired to recover bromine from the bittern, it is advisable to pass chlorine in the bittern at this stage rather than at the final stage. It is found that bromine which occurs originally in sea water or brine mainly as magnesium bromide, is gradually lost during the fifth, sixth and the seventh stages. This loss can be reduced to the lowest possible figure if high temperature is not used in the process and the mother liquor treated with chlorine in the sixth or the beginning of the seventh stage. This loss can be reduced to the lowest possible figure if thigh temperature is not used in the process and the mother liquor treated with chlorine in the sixth or the beginning of the seventh stage. Chlorine replaces bromine in the bromides and converts them in to chlorides. This operation is carried out in glass tower. Chlorine may be generated by any one of the well-known technical methods. After recovery of bromine the mother liquor contains mainly potassium chloride, traces of sodium chloride and magnesium chloride. This liquid if heated and concentrated to 40°Bé. Carnallite $KCl\ MgCl_2 6H_2O$ separates out.

Fig 5.10: Bittern Storage of 6th stage

Analysis and results of salt precipitated with chilling of bittern.

No.	Sample details are as under :		Density		Net sample weight
1	Liq.1	Feed bittern (source)	35.4	°Be	
2	Liq.2	26 ° C to 12 ° C	34.6	°Be	
3	Liq.3	12 ° C to 6 ° C	34.6	°Be	
4	Liq.4	26 ° C to 6 ° C	33.8	°Be	
5	Salt 1	26 ° C to 12 ° C			430.11 gms.
6	Salt 2	12 ° C to 6 ° C			113.86 gms.
7	Salt 3	26 ° C to 6 ° C			497.15 gms.

Note: Salt 1,2 & 3 received in slurry form. Filtered through vacuum pump for sample analysis.

	Salt 1	**Salt 2**	**Salt 3**
Volume in ml	200	60	202
Solid's weight in gram	169.49	35.56	226.77
Sp. Gr. (g/cc)	1.3183	1.3041	1.3209
° Be	35.01	33.81	35.23
Feed Bittern volume (ml)	1500	1000	1500
Total liquid remained (ml)	1370	970	1322
Difference (ml)	130	30	178

Parameters	**Salt 1**	**Salt 2**	**Salt 3**	**Liq.1**	**Liq.2**	**Liq.3**	**Liq.4**
Sp.gr.(g/cc)	-	-	-	1.3246	1.3066	1.2977	1.2997
° Be	-	-	-	35.53	34.02	33.26	33.44
	% As such	% As such	% As such	gpl	gpl	gpl	gpl
W.Ins.	0.00	0.00	0.00	-	-	-	-
Ca	0.00	0.00	0.00	0.00	0.00	0.00	0.00
Mg	9.31	9.02	9.23	99.27	95.50	95.38	96.23
Cl	15.25	17.2	17.73	273.04	269.50	269.50	269.50
SO_4	21.58	17.74	19.18	53.22	37.96	27.90	30.16
Bromine	0.00	0.00	0.00	4.13	2.69	2.94	4.90
KCl	8.07	6.89	8.73	20.54	13.54	10.04	9.75
Compositions	Salt	Salt	Salt	Bittern	Bittern	Bittern	Bittern
Parameters	% As such	% As such	% As such	gpl	gpl	gpl	gpl
$CaSO_4$	0.00	0.00	0.00	0.00	0.00	0.00	0.00
$MgSO_4$	27.05	22.23	24.04	66.70	47.58	34.97	37.80
$MgBr_2$	0.00	0.00	0.00	4.76	3.10	3.39	5.64
$MgCl_2$	15.06	17.74	17.13	333.78	335.01	344.38	344.30
KCl	8.07	6.89	8.73	20.54	13.54	10.04	9.75
NaCl	0.31	1.16	1.34	24.14	22.28	13.53	13.85
W.I.	0.00	0.00	0.00	-	-	-	-
Moit.(diff.)	49.51	51.97	48.76	-	-	-	-
TOTAL	100.00	100.00	100.00	-	-	-	-

Fig 5.11: Crystallized gypsum in condenser between 17°Bé to 25°Bé

6

Elements of Brine

Calcium Sulphate

It occurs widely distributed in Nature. As anhydrite, it is formed associated with limestone deposits and in salt layers. The dihydrite is more plentiful as Alabaster and Gypsum or Selenite. When calcium sulphate is precipitated from its solution as in the case of salt manufacture, it separates out as a dehydrate or gypsum. If gypsum heated to about 120°C to 130°C it loses one and a half molecules of water of crystallization and forms the semi hydrate $(CaSO_4)2H_2O$ known as burnt gypsum or Plaster of Paris. Gypsum is largely used in various industries.

When finely ground gypsum is heated with finely powdered coal or charcoal or with carbon monoxide, it is reduced to calcium sulphide and carbon dioxide. When calcium sulphide is treated with water, it is decomposed with the formation is calcium carbonate and sulphureted hydrogen which, in turn, may be ignited in air to give sulphur. These reaction are made use for making sulphuric acid at some location.

Calcium sulphate is very insoluble in water, at water, at 25°C, 0.0208 and at 30°C, 0.2096 grams are soluble in 100 ccs of solution. The effect of a foreign salt on solution is lower the vapour pressure of the solution at a given temperature. When gypsum is dissolved in the presence of sodium and magnesium chlorides, as is found to be the case in brines, the vapour pressure of the solution is distinctly lower than the vapour pressure of a solution of gypsum. In such cases a lower hydrate separates out. This explains the formation of anhydrites in the evaporation of brine. If the anhydrite forms on the surface and is in contact with moisture, it passes slowly in the dihydrite form or gypsum. The specific gravity of gypsum is 2.31. When plaster of paris is wetted with one-third of its weight of water, it forms a plastic mass which sets in form in 15 to 25 minutes to a white porous hard mass. This setting is well known for making moulds and in plaster work. The slight expansion which occurs during the setting of the plaster enable us to make sharp reproduction of the details of the mould.

Sodium Chloride

Generally known as common salt. It is very widely distributed in Nature as rock salt as Halite or dissolved in salt lakes as brine under the surface and in sea water. Pure salt is color-less and transparent. Salt crystallize in cubes. These crystals are anhydrous. A little sea water is mechanically held in crystals. This causes the salt to decrepitate when the heated. Sodium chloride melts 810° C and sublimes at higher temperature. 36.5 parts of salt are soluble in 100 parts of water at 30° C. The saturated solution thus contains 25.74 percent of dissolved salts. Its solubility decreases in the presence of other more soluble salts, such as magnesium chloride, magnesium bromide, potassium chloride and magnesium sulphate. This lowering of solubility accounts for the salt separating at 24.5°Bé instead of 26°Bé in the recovery of salt from brine. Its hardness is 2.5, its specific gravity ranges from 2.1 to 2.6 that of pure crystals being 2.135. In storage ¼ to ½ inch cubes take35 cubic feet per ton or 62.5 lbs per cubic feet. Smaller crystals require 33.4 cubic feet ton or about 67lbs. per cubic feet. The weight of salt in a cubic foot will depend upon the nature of the crystals and impurities contained. The rock salt of Mayo mines weighs 93 lbs. to cubic foot. In another case a compact natural block gave 131.44 lbs. Well-formed dry salt weight 60lbs.

Magnesium Sulphate:

Magnesium sulphate of commerce is chiefly obtained from kieserite ($MgSO_4\ H_2O$). At Stassfurt kieserite layer is found between carnallite and rock salt strata. It can also be prepared from magnesite and sulphuric acid. It occurs as kieserite in combination with potassium sulphate and magnesium chloride and contains six molecules of water. Magnesium sulphate is also obtained from dolomite – a mineral containing magnesium and calcium carbonate. If kieserite is dissolved in water, it takes seven molecules of water and forms a colourless crystalline salt readily soluble in water. This salt is generally known as Epsom Salt. It is efflorescent in dry air and begins to lose its water of crystallization at 150°C. 30.9 parts of anhydrous salt are soluble at 10°C in 100 parts of water. At 40°C 45.6 parts of water are soluble. If sodium chloride is obtained in the solution with magnesium sulphate, a double decomposition occurs on cooling the solution. Two molecules of sodium chloride with one molecule of magnesium sulphate give one molecule of sodium sulphate and one molecule of magnesium chloride. This reaction is made use of in the commercial preparation of sodium sulphate.

Magnesium sulphate is generally placed on the market as colour less crystal of 1.6787 sp. gr.

Epsom salt or magnesium sulphate is used in the tanning and dyeing industries, and in the manufacture of paints and soaps; but its chief use is in the finishing of cotton goods. Magnesium sulphate for finishing purpose must not contain any magnesium chloride, as the latter is liable to generate hydrochloric acid in the free state at the temperature of the callender rollers and thus tender the fiber. Epsom salt is also used for agriculture purpose, as weighting ingredient for sizing yarns. It is also used for weighting paper, silk and leather and for fire proofing materials.

Sodium Sulphate.

This compound in its anhydrous state is known as salt cake. Glauber's salt contains 10 molecules of water of crystallization. If Glauber's salt is allowed to remain exposed to the air or more quickly heated, it passes into the anhydrous form; the commercial variety is the decahydrate – Na_2SO_4 $10H_2O$. This salt occurs in nature in sea water natural brine and in several deposits in Persia, Spain, Caspian Sea and Siberia. For industrial purposes this salt is prepared from crude carnallite and also by decomposition of salt with sulphuric acid. Most of the Gluaber's salt manufactured is converted into soda. It is also found as by-product in the manufacture of nitric acid. Glauber's salt forms large monoclinic crystals which melt at 32.4°C. Glauber's salt is largely used by dyers, both in its crystallised and calcined form. It is used in the finishing cotton goods and is found in most of dyestuffs which are sold in the form of powders or pastes in which it is very often added for reducing the strength of the colours. It is used by dyers for purpose of regulating dying operations. Glauber's salt is principally used in the manufacture of sodium carbonate, and is also expensively used in glass making, especially window and bottle glass.

Potassium Chloride

This salt occurs in sea water brine and in various natural deposits. It was found in Stassfurt, ins. Salt Lake – Uta and at various other places, mainly associated with rock salt or brines. It occurs as seybine - KCl, as carnallite – $KClMgCl_26H_2O$ as kainite – $MgSO_4$ $K_2SO_4MgCl_26H_2O$, as schonite – $MgSO_4K_2SO_46H_2O$. These potassium salts are used for the recovery of pure potassium chlorides. Different methods were employed at Stassfurt to recover the pure salt. As potassium chloride and magnesium sulphate are equally soluble in water, the separation of potassium chloride is very difficult. Carnallite is, however, very soluble in hot solution of magnesium chloride. Magnesium sulphate and sodium chloride, however are sparingly soluble. The process of separation is based on this property of carnallite. In France the separation is effected by cooling the mixture of magnesium sulphate potassium chloride, magnesium chloride and sodium chloride. Magnesium sulphate and common salt separate out leaving the carnallite in solution.

Carnallite is separated out and treated as under:

1. Lixiviation of the carnallite with hot water.
2. Crystallizing the chloride of potassium by artificial freezing.
3. Evaporating and cooling the mother liquor to produce a second crop.
4. Treating the second crop as before.

Potassium chloride is a white substance crystallizing in cubes melting at about 750°C and slightly volatile at high temperature. 100 parts of water dissolve 32.7 parts at 15°C. Potassium Chloride forms the basis for the preparation of a number of other valuable potassium salts, such as potassium nitrate, carbonate, bromide, chromates and chlorate. By far the most important use of potassium salts is in the preparation of artificial manures. During the war the supply of potassium salts from Stassfurt was suddenly stopped and various other source of potash salt were brought to light. Potash salt are now prepared by chemical process or as a by – product in cement, salt and sugar industries.

Magnesium Chloride

Magnesium chloride is present in sea water in natural brine. It forms part of carnallite which is found in abundance in the natural deposits. Magnesium chloride of commerce is principally prepared as a by-product at salt mine and sea salt manufacture. In both case the mother liquor remaining after the recovery of the potash salts is evaporated at 134°C to specific gravity of 1.34. The hot concentrated liquor is run into casks or drums where it solidifies into a white mass containing about 45 percent. of magnesium chloride.

Calcium Chloride

Calcium chloride is present in sea water and natural brine. In the spontaneous evaporation of the mother liquor it separates out as tach-hydrite $CaCl_2MgCl_2\ 12H_2O$. There is no extensive application of this salt but as it is a by-product in several important industries, viz., in soda-ash manufacture and potassium chloride and the Welden chlorine process various uses are suggested for its disposal. Calcium chloride is a highly hygroscopic substance, which when dissolved in water shows considerable diminution in temperature. If a hydrated salt is heated strongly it becomes anhydrous. It is placed on the market as a fused solid. Its principal uses are as a dehydrating agent, preservative and in fire proofing paints. It is also used in freezing mixtures and in the manufactures of ammonium chloride and mineral waters.

Magnesium Bromide

Magnesium bromide is found in mother liquors remaining after the recovery of salt. Magnesium bromide as such is of little technical value except as raw material for bromine. As Magnesium bromide is highly hygroscopic, it is not separated by fractional crystallization, but by treatment with chlorine gas bromine is set free. Liquid Bromine is a dark reddish fuming liquid with a boiling point of 58.5⊠ Celsius. Liquid bromine has a vapour density equivalent to 3.1 times heavier than air. The threshold limit value (TLV) of liquid bromine is 0.1 ppm. It is widely used in the manufacture of Bromides (Bromine derivatives such as NaBr , HBr, MgBr etc which in turn are used in the manufacture of Dyes ,Medicines (In batch operations) ,Pesticides & Insecticides. Bromine manufacturing technology is divided into 2 forms i) Cold Process ii) Hot Process.

Bromine compounds are largely used in the production of 1,2-dibromoethane (ethylene di bromide), a lead scavenger used in making petrol antiknock compounds. Lead in petrol, however, is presently being phased out in many countries. This will clearly affect future production of bromine. It was once used in large quantities to make a compound that removed lead compound build up in engines burning leaded gasoline. Now it is primarily used in dyes, disinfectants, and photographic chemicals, fumigants, Flame proofing agents, water purification compounds, medicines, inorganic bromides (AgBr) used for photography, pesticides, water purification, used to make plastics flame retardant.

The basic Raw Material for Bromine production is Bittern, Liquid Chlorine, Caustic soda, Hydrochloric acid (32%). Bittern is a highly concentrated NaCl solution with a $MgBr_2$ concentration of around 2.0 gpl. Bittern is supplied from salt crystallizer's pond to the cold process plant Bittern has a density of 28° Bé and its concentration is achieved through solar evaporation.

Anhydrous liquid chlorine is basically meant for reacting with the Bromide of Bittern and liberate free bromine. 32% caustic is meant for the absorption of liberated free Br_2 vapour in it in order to prepare charge liquor with a concentration of 300 gpl(NaBr). HCl (32%) is meant for maintaining the pH of feed bittern in the order of 3 - 3.5 in order to facilitate the stripping process of Br_2 vapour.

7

Salt Production

Solar salt which relies on evaporation of brine in salt pans utilizing solar energy is eco-friendly. Sea water and sub soil brine, in fact, constitute a multi component salt system with number of salts dissolved in them. The basic operation of a solar salt field incorporates the progressive evaporation and concentration of brine in specially designed solar pans.

During the process of evaporation of brines a series of salts crystallizers out in the concentration ponds, the process being governed by the solubility products of the individual salts dissolved in it. Sodium chloride crystallizes out in the density range of 25°Be to 29°Be (specific gravity 1.209 to 1.250). Natural salt crystallizing out from brine contains impurities originating from brine compositions, the major impurities being calcium, magnesium and sulphate. The minor impurities that can pose problems for its applications are clayey matter, heavy metals, potassium, bromides, iodide and organic matter in the form of microalgae. Moisture in salt, mainly due to the presence of excess impurities of Magnesium salts, is also detrimental as it can lead to the caking up of salt besides being an unwanted baggage during transportation of salt. For example, if 10 million ton salt has to be transported from the fields by road, a 4% moisture level in salt would amount to 40,000 ton un-necessary trips! A dry salt thus helps reduces the carbon footprint.

The impact of impurities in salt depends on its application. Bureau of Indian Standards (BIS) has recommended certain specifications for industrial grade salt.

Table 7.1: The BIS Specification for industrial grade common salt

Constituents		Industrial Grade - I	Industrial Grade II
Sodium Chloride (as NaCl)	% by mass, Min	99.50	98.5
Matter insoluble in water	% by mass, Max	0.05	0.2
Calcim salts (as Ca)	% by mass, Max	0.03	0.2
Magnesium salts (as Mg)	% by mass, Max	0.01	0.1
Sulphate (as SO_4)	% by mass, Max	0.20	0.5
Iron coumpound (as Fe)	Parts per million, Max	10.00	20.0
Trace metal impurities		Within Permissible limits	

In addition to the above specifications, some chlor-alkali industries insist on the limiting levels of iodides, bromides, potassium and other trace metal impurities as they have to invest considerably in eliminating these impurities. For example traces of iodine in salt beyond the prescribed limits detrimental for chlor-alkali manufacture, whereas iodine is an essential micronutrients for the human body necessitating iodization of edible salt. Similarly, as will be seen from the desired industrial salt specification of a leading company.

Table 7.2: Specification of salt desired a leading chlor-alkali industries

Parameters		Percent (%)
Calcium salts (as Ca)	Percent by mass, Max.	0.04
Magnesium salts (as Mg)	Percent by mass, Max.	0.02
Sulphate (as SO_4)	Percent by mass, Max.	0.11
Iron (as Fe_2O_3)	in ppm	15.00
Chloride (Cl)	Percent by mass, Min.	59.01
Sodium (Na)	Percent by mass, Min.	38.25
Potassium (K)	Percent by mass, Max.	0.01
Insolubles	Percent by mass, Max.	0.01
Moisture (H_2O)	Percent by mass, Max.	2.50
Sodium Cloride (NaCl)	Percent by mass, Min.	97.22

Potassium level is specified as <0.01% (100 ppm), whereas salt with high potassium level (5-50 %) are sold as low sodium salt.

Mechanization and Modernization of Salts Works

The basic operation of solar salt field involves the progressive concentration of seawater in a series of large open ponds called reservoirs and condensers, followed by further brine concentration ponds saturated with respect to Sodium Chloride in a series of ponds called Crystallizers. Salt gets precipitated in these ponds.

The Crystallized salt is harvested from these ponds and transported to Washery where Salt is usually upgraded by various washing processes which ensure the final product meeting customer specifications and requirement.

World demand for salt currently exceeds 235 million tons. Around 40 percent of salt production is by the solar evaporation of seawater. Until the 1960s, China, India and US dominated the production of solar salt worldwide, and they still remain major producers.

Historically, productivity and quality of solar salt produced in China and India were low. As alternative supplies were unavailable customers using this salt for both edible and chemical production purpose accepted this quality.

The rapid growth of Japanese Chlor Alkali Industry during the 1960s and 70s, using new high cost technologies, called for high quality salt competitively priced. This demand could not be meet from Japanese domestic supplies. New Solar Salt production facilities were therefore set up in Mexico and Australia to meet this demand supported by Japanese companies.

Supply of Salt to Japan is mainly dominated by Mexico and Australia, who implemented latest Salt Production Technology. They have progressively increase production to meet demand from new Chemical customers in South Korea, Taiwan and Indonesia. Prior to the advent of the Solar Salt fields in Mexico and Australia, Solar Salt from China and India was exported in limited quantities to Japan and other Asian countries. Such exports, now only occur during brief periods when supply from Australia or Mexico are disrupted due to adverse weather conditions. The lack of surplus Solar Salt worldwide has forced China to revert to vacuum salt utilizing rock salt deposits. This process is extremely energy-intensive compared to the free use of solar energy in the production of solar salt. All Chlor Alkali plants in China are now being serviced with Solar and mine Salt production within the country. The increased demand for salt in China from the growing alkali industries has also necessitated the use of high quality as ion exchange membrane is the principle technology employed by new chlor alkali plants.

Reverting to Solar Salt, despite uncontrollable external factors such as variations in climatic conditions, seepage of brine etc., technology as now improved to overcome these difficulties and more effective designing of new fields and efficient operations have come to be adopted. One of the major improvements is to precisely control the movement of large volumes of brine by use of continuous and correct brine mass balance over the field. Such models are now available by using evaporation rates, rainfall, seawater quality and seepage losses from any given area. This will give us the estimation of average annual salt production. Alternatively if several sites are under consideration, the design will determine which site will best achieve a higher production rate. Further optimization steps may be made incorporating economic and operational criteria. The number and orientation of concentration ponds will be optimized so as to reduce the initial capital costs for construction of ponds levels, pumping stations, transfer weirs etc., whilst at the same time ensuring that the final layout will not compromise the efficient operational control of brine hydraulic, chemistry and biology. The layout of crystallizer ponds will be calculate keeping in view the topography of the land, harvesting period, weather conditions and downtimes etc. Following such calculation improved methodology can be developed even estimating month-wise production, using average monthly conditions. Salt field operations therefore should be scientific and control of the Salt field operation should be as per the prevailing actual change in the climatic and other conditions. Studying the month wise mass balance throughout the salt field should become an important factor for controlling the intake pumping, charging to the crystallizers and discharging bitterns. Control of density of various streams of brine would very much help the output and the quality of the Salt produced. It would drastically minimized the losses which occur at several stage of the process.

The benefits accruing from these methods include reduction in capital costs in establishing the field and in the reduction of operating costs and enables the operational staff to closely monitor brine flow Throughout the field so as to achieve both chemical and biological control. This ensures that the quality of salt deposited in crystallizer ponds in predictable in both series and parallel flow systems. Washing plants can therefore be designed to minimize capital and operating costs for production of high quality salt. Indian can definitely compete in the world market successfully by implementing these new techniques of developing salt works very large scale.

The mechanization and modernization not only relates to harvesting methods but also relates to pumping operations, brine flow management and concentration controls with biological management in the circuit as well as processing of raw harvested salt to finished product. It is well known that solar salt works are most efficient converters of solar energy into an inorganic commodity. Conversion rate of solar radiation and removal of water vapor from the brine to the open atmosphere takes place with record efficiency. Only brine pumping, salt harvesting, processing and conveying involve consumption of electric power. The energy balance shows that solar salt requires only a fraction of manmade energy compared with salt produced by solution mining and vacuum evaporation with employment of advanced technologies for biological management and brine control in solar salt works, harvesting techniques and salt processing, it is possible to produce salt of very high quality with very low losses. Elimination of back-mixing of brines and gradual change of brine concentration improve salt crystallization rate and salt quality, effectively contributing to salt production economy. Modern harvesting equipment operated under precise level control is capable of near 100% recovery of salt from the crystallizer floor with negligible salt contamination. Advanced salt purification technology is able to completely remove the mother liquor from the salt crystals and also minimizes washing losses.

Pumping and Density Control:

Pumping requires careful calculations of the volume of brine needed daily and this depends on the rate of evaporation of sea water / brine with respect to concentration, gradient and the brine levels in ponds. Such careful calculated flow of brine will control the quality of brine from pond to pond and right quality of feed brine to the salt crystallizers. Computerized programmed / mathematical model on brine movement to achieve desired result should prepare by each salt works.

At solar salt works mainly there are two type of pumps are in use. 1. Vertical turbine pump & 2. Mix flow pump. The capacity of each may vary. As observed Vertical turbine pump is the best suitable pump for Sea water pumping and brine transfer. The capacity depends on the requirement of flow, head and quantity. These type of pumps are generally vertical turbine pump with 75 hp prime mover rating flange mounted motor, 600 x 600 mm size, with 6 meter head. In these type of pump the biggest challenge is dry run. The maximum observed problem occurs in such pumps are dry run or low level pumping. To prevent such problem automation should introduce to start and stop pump with respect of pump's suction sump level. At present normally there are two type of sensors are using 1. Ultra sound sensor with PLC & 2. Level probe. Such system can operates automatically with respect to tide level and in condenser level of condenser. The operator only need to manage the duration of pumping as per requirement of brine supply.

Fig 7.1: MF pump with suction dome

For transfer of brine and bittern for long distance or at higher head Mixed flow pump or split case pump are preferable. In salt works split case pumps are better due to easy in maintenance. For salt production there are two type of provision is necessary. One is suction dome and another is sealing/flushing provision. In mixed-flow/split-case pump there is requirement of foot valve at suction line. But at salt operation there is observed frequently jamming and damaging of foot-valve, to prevent this there must be suction dome (Three time higher volume then suction line.). This can eliminate foot valve from suction line. Frequent jamming of salt observed in brine and bittern pumps, so if low density brine is available nearby then one small pump is preferable for washing of pumps and lines to prevent jamming and also for sealing water of pumps lubrication purpose.

Fig 7.2: Vertical turbine pump with auto operation sensors

Salt Harvesting

Mechanized harvesting calls for even bed levels in the crystallizers required bearing capacity of the crystallizer beds, optimum bed thickness and mode of movement of salt. At present we are harvesting salt by Harvesters or by excavators and moving it to wash plant or consumer in one or two stages. The economics of handling cost and the contamination levels of insoluble determine single or two stage movements, this can be modified by shifting to conveyor systems and unloading salt at the relevant dike to facilitate onward movement to the wash plant. The size and capacity of harvesters and handling equipment can be decided based on the capacity and size of crystallizers. It may be differ from one salt work to other. In fact while designing logistics it is very important to match the loading rate of harvesters with the time cycle and capacities of handling equipment with reference to the distance of the destination for getting optimum output from harvesters. This also requires one to keep in mind the handling capacity of the washery plant. The washery plant in itself should have to match with the individual salt works. If the capacity of washery plant is less than the salt production then we have to make salt stack to store salt. If the capacity is higher than it will get idle time. For supply to soda ash manufacturer supply of rain water or fresh water washed stake is also in practice. However, the size of salt stake of the product salt is equally important. It is essential to drain of any moisture attained during washing operations after giving some residence time. It is also important to consider the rate of dispatch. All these aspects determine the size, height and length of stackers; moveable and immovable. It also determines the logistics of dispatch machineries.

It is very clear that every operation / salt producing unit has its own unique design parameters. The deployment of harvesting machine and process to the stock pile should be carefully planned and simulated to achieve desired result of reduction of operating cost and minimizing investment cost.

Theory & Practice of Solar Salt Harvesting

The methods and machinery involved in harvesting and transporting solar salt from the crystallizers to a wash plant are as varied as the weather and soil conditions where the salt is produced. The amount of evaporation and rainfall including the way the rainfall occurs has a great effect on the selection of method to get the salt out of the pond without disturbing the mud below. It is very important to not mix the mud into the salt at the mud salt contact point and to not track bottom mud or road mud or gravel onto the pond salt surface to be picked up by the harvester equipment.

It is the nature of solar salt that the solid crystals of salt are laid down in large flat ponds and the salt has to be transported to the pond edge to get it into haul equipment to transport the salt to a central location for stacking or washing and further processing. The usual and the most common is to transport the salt over the surface of the undisturbed salt crop or over the salt's road in case of thin crust which transports later on. There have been some successful slurry transport systems-harvesting under water but not in any major size salt plants. There are slurry transport systems from a crystallizing area dump pit to a central wash plant in the US and China.

Salt Floor Harvesting

The simplest solution is the use of a salt floor of enough thickness to allow the use of standard highway type haul equipment and place salt on the surface of the roads to keep dirt from being tracked onto the pond surface. This is the system used in most of the world's major salt plants supplying the chemical industries that are using solar salt.

There are variations of the salt floor option where the climate is marginal and part of the salt crop is left in the pond occasionally to re-grow the floor depth.

There are many types of systems used with salt floors, probably the simplest is to just rip the crop and grade it into a large windrow and load it into trucks with Loaders or Excavators. Some of them double handling the salt to keep heavy equipment off the pond surface and to keep road impurities from being tracked into the pond.

Fig 7.3: Salt Floor Harvesting

One of the harvest systems most widely used in large solar plants in the Palmer type machine that has evolved from the unit first developed near the Great Salt Lake in the US in the early 1960's. There is one unit of the basic design rated at 3000 tons per hour at Essa in Mexico and serveral others rated at 500 tph., 750tph and 1200 tph. These are currently designed and built by J Frank Bonnell of Salt Lake City. A series of smaller harvesters, primarily for salt floor use are built by Serra Machinery of Spain to load various types of trucks, trailers and conveyors. Both of these type machine use a scoop blade at the front bottom with a drag conveyor right behind the blade to elevate the salt up a slope and discharge to a cross rubber belt conveyor slopping up over the haul truck or trailer transport equipment. These two principles are used for just about all harvesting machines, the use of drag conveyors at steep slopes right from the pickup point to keep the machine length short and to elevate the salt quickly.

The original depth of cut control for the Plamer design machines was to insert the scoop like pick blade into a 'split' in the salt crop.

The 'split' was a fracture plane in the salt at the base of the current years salt crop. The split was created by entering the pond with brine in it after making a small amount of salt and dragging the salt surface with a long piece of angle iron behind a farm tractor to create a layer of loose salt crystals-the 'Split'. The harvester pick up blade was simply shoved down until it reached the split-the new crop would lift slightly ahead of the blade and the forward travel speed could be increased. The lift rams were set at that point until a pass down the pond was completed.

As this harvesting machine was used in areas where salt depths of greater than 6" salt crops were common, the salt was ripped with a grader and pulled up into a single windrow for the harvester. After the harvester had passed; the grader flat the area-there is enough loose salt to create the spit which fractures when the graders rip the next crop.

This machine also uses laser guided machine blade control technology to establish the cut depth in some plants today and there is no concern about split. Various types of skids at the front of harvesters sliding on the top of the salt crop are used to provide an adjustable gauge to regulate the depth of the cut.

Fig 7.4: Laser guided salt harvester with Precision cut attachment at Shark Bay – Western Australia

Excavators:

Several types of road building machinery are used to pick up and load out salt. The most common of these are excavators, tractors and trippers. This type of machinery are using in medium scale salt works and where the salt bed is lesser then 12 inch. To harvest this salt, in first stage hydraulic excavators lifts salt and collect in crystallizer to make salt road on the salt bed. And in second stage the same salt loaded with excavators. This type of operation needs excavator operator's higher attention for salt harvesting otherwise it will damage salt bed. In some case in absence of salt harvester we can use excavators and dumpers or trucks to lift and transport salt from salt crystallizers. In some case small tractors with 3 to 10 mt capacity are also using. For all above operation we need to use plane edge bucket / shovel, if we use toothed bucket / shovel then there may be possibilities to damage and disturb salt bed. All above damaged salt bed is the big reason of insoluble in harvested salt. And it will also make difficult to make even layer for next crop. This method is very cheaper compare to harvester operation for medium scale salt works. Some saltworks are using wheel loaders, front end blade tractors or dozers. All of these equipments are not suitable for salt harvesting operations. Because all of these equipments are having back side counter weight and back wheel have higher ground pressure, if equipment works on crawler then the chain forced towards the back side and all weight transfer towards the back side of equipment and this is the cause of damage of salt bed.

Fig 7.5: Excavator Collecting salt to load truck.

Fig 7.6: Excavator loading collected (harvested) salt into tractor tripper.

Fig 7.7: Salt in windrow awaiting for loading.

Graders:

The grader is the good option for salt harvesting operation. Grader can also work with hydraulic excavators, in such type of operation the grader rip out salt on salt crystallizer's surface with making salt's small embankment that can load by excavators with plane edge bucket / shovel. Grader can also equip with laser guide reference and it can provide regulated depth control facility. Graders are very fuel efficient equipment compare to excavators so, for ripping of salt more preferable is to use motor grader in place of excavator. This ripped salt can load in trucks. Grader comes with wide variety of different type of attachment, in hard surface the operator can use ripper attachment to losing the salt surface, then slide the ripped salt to collect the salt. 14 feet blade grader is more economical for salt operation, yes it is heavier in weight but can drift more salt and also have higher power output. The biggest benefit of grader operation is the precision. We can get high quality of depth control than excavators, if machine equipped with laser guidance then the depth control can be in millimeters.

Fig 7.8: Grader Operation wiping salt.

Fig 7.9: Grader Operation wiping salt to load in tripper through harvester or excavator.

Surface Mining:

Some large salt works are using surface mining technology with using modern surface miner. This is suitable in thicker salt bed and large crystallizers. This type of technology is using in Australia, India, Iran and Turkey.

These machines are compact and efficient in work. These equipment equipped with milling drums to crushing and harvesting salt quickly and in demandable size. The cutting tools and milling drums are fact and simple to replace. These surface miners comes with loading conveyor to load salt to trucks. Surface miners slices salt layer to get required salt surface for quality control. Some machines comes with laser guided leveling reference, so operation is automated and can achieve level accuracy in millimeter. Only drawback in these machines is, these machines are very heavy and cannot operate in thin layer of salt.

Fig 7.10: Salt harvesting with surface miner in Onslow, Western Australia.

Fig 7.11: Salt harvester with surface mining technology. At Gandhidham, Western India.

Mud Floor Harvesting

Most of the salt operations scattered all over the world are in the climates with too much rain. The salt in these operations are usually made in thin crops and taken out to avoid rain. In such systems the salt which precipitated in the crystallizer must harvest before rain. The heavy rain usually washes all remain salt and not available for next season for primary bed. In this pattern whole salt till soil layer should lift either though salt harvester or manual operation. Though the layer is so thin and not enough to hold weight of trailer, or any transport vehicle, generally mobile conveyors are in use with salt harvester. Salt harvester harvest salt and dump on mobile conveyor. These mobile conveyors, shift salt to desired location, either on salt stake or on silo hopper to load tripper trucks.

Fig 7.12: Mud Floor harvesting

Fig 7.13: Harvesting operation with Salt harvester, Mobile conveyer, Excavator. At TATA Chemicals Ltd. Mithapur, Western India

Fig 7.14: Salt harvesting in action. TATA Chemicals Ltd. Mithapur. Western India.

Fig 7.15: Mud floor Salt Harvesting with Mobile conveyor & Excavator.

Manual Salt Harvesting:

This is the most common and oldest method of salt harvesting, and using in most part of the world in these days also. These operations are for the most part too small to undertake special equipment steps to mechanize and there has been no mechanical system. When the crystallizers are too small and the salt bed is thin in such type of crystallizers people are favoring manual harvesting. Their tools and gears are different in different part of world. In India people uses raking while the crystallizer charged. This also helps to loose the salt crystals and to increase salt precipitation with increasing surface area of salt crystals. While racking done the surface area of salt crystals increase so the contact area of salt crystals increase will increase the crystallization, as well as this increase the wave movement of water surface which increase the evaporation rate of brine.

Fig 7.16: Raking activity

This loose salt collected in the crystallizers with wide wooden shovel, lifted and transfer to salt platform which is located parallel to salt crystallizers. This is complete manual process which is very cheap in cost as well as good for very thin layer of salt in terms of soil (insoluble impurities). The rate of production is less but while the crystallizers are approximately 1 acre in size then this method is highly suitable.

Fig 7.17: Manual Salt Collection

Fig 7.18: Manual Salt lifting & Shifting to stacking platform

Mechanical Salt Washery

Solar Salt can be obtained at a rather good quality depending upon management of crystallization ponds: when harvested, special care is taken to avoid contamination by soil and other impurities. Some impurities can be trapped inside the crystals or between the crystals. Washing is done to improve the quality. In some case, intense mixing or mild grinding to free some of these impurities can precede washing. Washing or washing plus hydraulic classification can separated these impurities. Harvested salt fed with various brine consists mainly of small and hopper-shaped crystals whose cavities entrap contaminates similar to brine mother liquor, prove detrimental in upgrading the quality at wash stage naturally or artificially.

Fig 7.19: Salt Washery with twin classifier

Common salt as produced by solar evaporation of sea - water or sub-soil brines through conventional process is used for industrial applications, wherein highly reduced levels of Ca, Mg, SO_4 and insoluble impurities are preferred. These impurities are removed by resorting to mechanical washing of salt. Various types of washeries are in use across the world. Hydrocyclone and screw type washeries are most commonly used.

It has been observed that in the case of sea - salt nearly 70 % of Ca is present in the superficial form and in the case of sub - soil brine, > 70 % of Ca impurities are embedded in the salt crystals. In the case of Mg impurities, about 80 - 85 % are present in the superficial form while 15 - 20 % are embedded in the agglomerated salt crystals. It is observed that the effective removal of impurities can be achieved from freshly harvested salt rather than the stored salt. Moreover, lesser the level of impurities in the feed salt, grater the efficiency of impurities removal.

Initially, salt is unloaded to Dump Hopper With a storage capacity varies depends on washery capacity. From the storage hopper one weigh feeder will feed the required capacity of salt to a classifire. The Classifire will remove all the impurities and insoluble, which are particularly light and rise towards the surface to pass into the draining tanks. Classified salt conveyed to the metallic conveyor for further washing for upgrading salt quality and using the same metallic conveyor for washing de-watering.

Fig 7.20: Double Classifier with twin screw conveyor

The salt wash involves operations as per the necessities to separate solid and soluble impurities from sodium chloride crystals by simple and separate physical means.

The soluble and solid impurities are subsequently separated by means of dissolution and hydro-mechanical classification steps. By addition of water the soluble impurities are washed out and purged with a liquid effluent stream, which may be taken to feed a solar pond to improve the total process yield.

By taking advantage of different particle size of sodium chloride crystals and solid impurities, the process contains steps with hydro-mechanical classification. The Gypsum crystals and other solid particles, like sand and clay, can be separated from the sodium chloride crystals. The separated solid impurities are concentrated and discharged.

In order to effect better washing of the Salt, most design improvements and modifications of operational characteristics are required to be carried out.

Fig 7.21: Metallic Conveyor

For e.g. the screw conveyers are to be installed with a suitable inclination to the horizontal, counter current flow of circulating brine is to be streamed in with a lower solid-liquid ratio of Salt per liter of liquor, optimum contact time of solid and liquid is to be established to minimize the loss if Salt suspended impurities of circulating brine are to be got rid of by adding a coagulating agent and allowing this to settle.

At the end metallic conveyor placed to reduce magnesium impurities to wash salt with low density brine. Salt spreads on the moving screw conveyor and sea water or further low density brine sprinklers on it for further purification. The final salt quality is depends on the density & quantity of sprinkled brine, speed of metallic and screw conveyor and metallic conveyor.

Fig 7.22: Salt washing on metallic screen conveyor

In general practice high concentrated brine is using in screw conveyers higher than 20°Bé. For metallic conveyor initially higher concentration like 20°Bé and in second stage low concentration like 0°Bé - 4°Bé as per output quality requirement from buyers. The discharge of metallic conveyor can collect and feed as input washing brine for screw conveyor. The discharge of screw conveyor contains high insoluble depends on the raw material so it should settle and can use to produce off grade salt after crystallization. The pressure and flow of higher density brine should be higher than low density brine. The low density brine's pressure and flow on the metallic conveyor would be lower like drip sprinkler to reduce Magnesium impurities from salt. The purpose to use metallic conveyor is to drip washed brine from salt after washing immediately.

If we have to produce salt for soda ash grade i.e. Calcium 0.20% & Magnesium 0.10% then washery with Screw conveyor (classifier) and metallic conveyor is not viable because in general the quantity loss from such washery reaches up to 20%. For such purpose washery with vibrator screen is preferable because less contact with washing brine and washing time is less. The output also become higher than other washery. Selection of washery must be depend on the requirement of finished material and raw material.

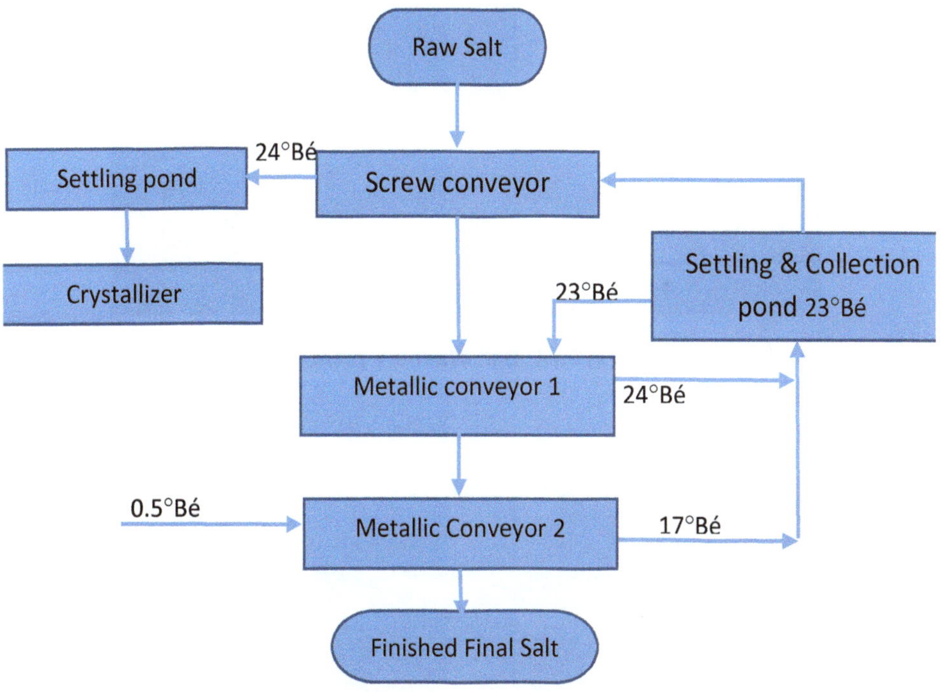

Fig 7.23: Salt Washery process flow diagram

Some analysis of washery brine and salt:

	Ca^{++}	Mg^{++}	°Bé	
Classifier Input brine	0.08	0.65	23	Stage I
Classifier output brine	0.10	0.65	24	Stage I
Metallic conveyer input brine 1	0.08	0.65	23	Stage II
Metallic conveyer output brine 1	0.12	0.65	24	Stage II
Metallic conveyer input brine 2	0.12	0.04	0.5	Stage III
Metallic conveyer output brine 2	0.10	0.16	17	Stage III
Raw Salt	0.29	0.29		
Classifier output	0.11	0.14		Stage I
1st washed, metallic conveyor	0.10	0.09		Stage II
2nd washed (final)	0.08	0.04		Stage III

Fig 7.24: Salt Washery plant

8

Scientific Approach in Salt Production

Salt Production by Series Feeding Method

The solar salt with less contamination of Ca can be produced by adopting the proper methods of brine management such as deep charging of crystallizers and adopting the method of series feeding for charging and discharging crystallizers. The result are given in table given below.

Though this process offered common salt with less impurities of calcium in first few fractions, the salt fraction separated at a later stage contained higher Mg & SO_4 impurities which are further purified by resorting to the mechanical washing. Moreover, the Ca impurities are not reduced to a level required by chlor-alkali industries.

Analysis of salt obtained from series and parallel feeding pans

Sr No	Salts	Parallel Feeding Pan	Series feeding Pan					
			S1	S2	S3	S4	S5	S6
1	Ca^{++}	0.16	0.18	0.15	0.11	0.11	0.09	0.08
2	Mg^{++}	0.21	0.14	0.16	0.19	0.27	0.35	0.44
3	$CaSO_4$	0.55	0.61	0.51	0.39	0.37	0.32	0.29
4	$MgSO_4$	0.35	0.23	0.26	0.32	0.46	0.58	0.34
5	$MgCl_2$	0.56	0.37	0.42	0.52	0.74	0.91	1.17
6	NaCl	98.3	98.3	98.3	98.7	98.3	98.1	97.8
7	Concentration Range °Bé	25.5 to 30.00	25.5 to 26.00	25.8 to 26.30	26.3 to 27.00	26.50 to 27.5	27.00 to 28.00	27.5 to 30.00

Production of solar salt from brines employing the principle of common ion effect.

In further development the reduction of Ca in solar salt was achieved by employing a common ion effect in the brine systems through addition of either $MgSO_4$ (Bittern) or Na_2SO_4 in the saturated brine. The recycling of bittern containing higher amounts of sulphate is being commonly practiced by the solar salt producers. This process ended up with the harvesting of common salt with slightly reduced levels of Ca impurities but the Mg and SO_4 impurities were found to be on the higher side.

Specifications of salt produced by the addition of $MgSO_4$ (Bittern) & Na_2SO_4

Description	**Sub soil Brine +$MgSO_4$ /Bittern**	**Brine + Na_2SO_4 / effluent containing Na_2SO_4**
Constituent	% w/w	% w/w
Ca	0.17	0.17
Mg	0.32	0.24
SO_4	0.84	0.77
NaCl	98.01	98.36

Production of solar salt from brines using polysaccharides and marine microbes

In a subsequent development Ca levels in sea salt was reduced with addition of very low levels of certain polysaccharide in the concentrated brines. This process offers common salt with reduced level of calcium but magnesium and sulfate impurities are no reduced. In other process, scientist identified certain types of marine microbes which can mop up Ca ions from brines and reduce Ca impurities in the crystallized salt. They succeeded in developing a novel process for producing salt with comparatively lower levels of calcium. However, the Mg & SO_4 impurities could not be reduced by this process. Further details will be on next chapter of Biological Management.

High purity salt through de-sulphatation process

It is lower concentration of sulfate present in subsoil brine responsible for deteriorating the salt quality. If sulfate from brine is removed through forced intervention prior to the crystallization of salt, not only good quality can be produced but the bittern generated can be directly processed for the recovery of valuable marine chemicals. The process uses inexpensive source of $CaCl_2$ such as distiller waste of soda ash plants as a desulphating agent. Subsequently, a process for the recovery of common salt and marine chemicals from brine in an integrated manner was developed with (US patent No. 6, 776, 972 dated on 17th August, 2004). To optimize the addition of calcium chloride to brine for effective desulphatation, we can determine the solubility behavior of $CaSO_4 2H_2O$ in aqueous NaCl solutions up to concentration of 25% NaCl (w/v).

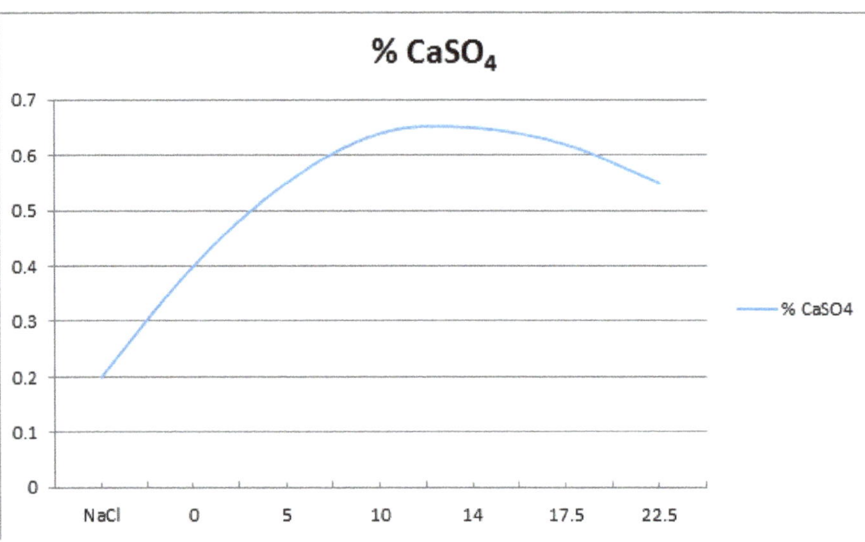

Fig 8.1: Change in solubility of $CaSO_4$ at varying concentration of NaCl in system of both the Salts.

Addition of Calcium of chloride reduces the solubility of $CaSO_42H_2O$ quite dramatically while maintaining the basic pattern of the solubility curve. The de-sulphatation process has been optimized by understanding of ionic interactions in brine systems and mapping the solubility pattern of gypsum at various salinity levels in presence of different ions in sea water systems. Different ionic processes like ion solvation, ionic relaxation effects and ion pairing occurring in brine systems have been systematically investigated for optimizing de-sulphatation process. Solubility pattern of gypsum in presence of different salts as a function of salinity.

By manipulating the composition of brine by forced intervention, co-precipitation of gypsum with sodium chloride can be prevented by suppressing its solubility in brine at initial stages itself.

Salt produced through de-sulphatation process (washed in salt washery) as analyzed by Tata Chemicals Limited.

Parameters	% w/w
Calcium	0.024
Magnesium	0.029
Cl as NaCl	99.36
Sulphate as SO_4	0.125
Moisture	0.039
Alkalinity as Na_2CO_3	0.079
Iron as Fe_2O_3 (in ppm)	11.44

The specification of salt produced from subsoil brines of Little Rann of Kutch region through de-sulphatation.

Constituents	% w/w
Ca	0.07
Mg	0.06
SO_4	0.15
NaCl (on dry basis)	99.10
Insoluble	0.06
Moisture	1.20

The de-sulphatation process works well for any kind of brine and can be carried out on a large scale. The process is most economical in solar salt works having the availability of destiller waste liquor source containing calcium chloride in the vicinity.

Effect of high Magnesium in salt crystallizers.

In some case while high magnesium in brine due to no replacement of feed brine, mixed brine with bittern to enrich Be or due to any other case the magnesium level of soil will increase and the soil of crystallizer become soft like sludge and become dark black due to decomposition and change in structure of soil particles. However the salt bed may thick but it cannot prevent to float up soil particle. To improve crystallizer we have to wash slightly crystallizer with shallow charged low density brine (22 be to 23 be) at time of discharge of bittern. This will replace $MgSO_4$ with $CaSO_4$ from soil particle. This may wash some amount of salt but will increase the quality of salt and improve salt bed. Should repeat such wash depends on the condition of crystallizer. This process is useful in such crystallizers which are not seasonal and not removing complete salt before rain.

Fig 8.2: Sprinkler arrangement on pump delivery line to enhance evaporation rate.

Fig 8.3: Crystallizer with highly concentrated planktonic *Halobacterium salinarium*

9

Biological Management for the Salt Manufacturing

Technology and guidelines for site selection, system planning and construction, and management of solar salt works are available from engineering firms, governmental agencies, private and international organizations, trade journals, other publications and person's experiences in the management of solar salt works. The resulting solar saltworks, designed, constructed and managed with traditional concepts from the above sources, have an initial period of successful manufacture of sodium chloride (NaCl). After the initial period, from the outset and usually lasting for 7 to 12 years, unforeseen events occur which require considerable expenditures to maintain quality and quantity of the manufacture salt.

Events first observed includes: (1) Excessive loses of salt in the wash process requires to reach world standard assays, (2) Soft crops and salt floors in crystallizers unable to fully support harvesting machinery, (3) Decreased quantity of harvested salt, and (4) Prolonged residence of salt on the stockpile to improve quality.

Fig 9.1: A thin dark black layer of biological substance occurred in old crystallizers mixed with salt while harvesting. Generally disappears and de-colorized while exposed in sun light.

Fig 9.2: Clearly visible different layers of salt, soil (deposits while in rainy season) & organic matter.

Subsequent events include: (1) Change of salt crystals from mostly single, large, transparent and solid to conglomerates of small, opaque hoppers with layers and cavities that trapped and bind contaminants, (2) Deposition of most gypsum as microscopic crystals, (3) Accumulating organic and inorganic substance in concentrating ponds with losses of volumes and surface areas, and (4) crystallizers with viscous brine.

Remedial measure:

Pond modification, wash machinery changes, brine handling alterations and adjustments of harvest techniques where often short lived, marginally effective, and expensive. These undesirable events and remedies are the result of outdated construction and management procedures coupled with inability to manage the biological system, and failure to deal with increasing nutrients in the saltworks.

All above are the causes of the reason of the Biology, The neglected technology for the manufacture of solar salt.

Desired Biological System in Solar Saltworks with Seawater Intake

A biological system able to help or harm salt manufacture inevitably develops in every solar salt works. Each biological system consists of (1) Producer micro-organisms: Photosynthetic algae, Cyanobacteria, and bacteria that use light, carbon dioxide and minerals to manufacture new organic substances, and (2) consumer organisms: Bacteria, crustaceans, mollusks, and protozoa that consume (eat, oxidize) organic matter. In each pond microorganisms are organized in a suspended community and a floor community. Suspended microorganisms are the chief producers of new organic substances, they move downstream with the water flow and their colour traps solar energy, heats the water, and aids evaporation. Ponds of low salinity (3.5°Bé to 9.0°Bé) occupy the largest surface area and volume of saltworks; microorganisms in this range manufacture the largest quantity of organic substances, they consist of the greatest number of species, and power the entire biological system of the saltworks at desired levels. After the low salinity microorganisms reach middle salinities (10°Bé to 18°Bé) they die and disintegrate, and their released substance become nutrients for a new set of microorganisms with fewer species than in the previous ponds; microorganisms of intermediate salinity range consume more organic substance than they manufacture by photosynthesis. As the intermediate salinity microorganisms flows into the high salinity ponds (19°Bé to 30°Bé), they also die, and their released substances become nutrients for the few species of algae and bacteria of this concentration range. Macroscopic consumer organisms copepods and mollusks in low salinity Artemia in middle salinity that ingest unicellular algae, protozoa, and particulates release fecal pellets, clear the water, and remove nutrients. In high salinity red halophilic bacteria are the chief consumers of organic substances.

Floor communities of desired biological systems, composed of producer and consumer microorganisms organized in firm, layered mats in most ponds controls leakage, maintain desired thickness, remove nutrients from the overlaying water, stay in place, and contribute to the delivery of high quality brine to the crystallizers.

Technology for Obtaining and Maintaining a Desired Biological System:

In new solar saltworks desired biological systems are best accomplished at the planning and construction stage, but in existing solar saltworks favorable systems can also result from appropriate pond modifications, equipment improvements and proper management procedures.

Establishment and maintenance of desired biological systems required stable physical conditions in every concentrating pond: unchanging salinity at each point, uniform depth and flow rates, and moderate wind and wave action. Stable conditions over appropriate distances and time in each pond enable maintenance of large numbers of important species of microorganisms in both communities. Numerous species result in production of microorganisms adapted to narrow salinity ranges able to perform many useful functions, compete for resources (nutrients, light, food) and prevent domination of undesirable microorganisms, such as the inedible Ochromonas in low salinity, the mucilage producing Aphanothecehalophytica in intermediate salinity, and the organic releasing Dunaliellasalina in high salinity; these three often shade out important microorganisms, increase viscosity of water, and prevent development of useful (nutrient-removing) floor communities, and strongly bind contaminants.

Physical requirements for stable conditions include sufficient numbers of concentrating ponds each with a small salinity range to minimize adverse effects of any back mixing on the microorganisms, use of proper calculations and mathematical techniques to control depths, flows and salinity, adequate numbers of fully adjustable gates, auto level guided auto pumps and standby pumps at each pump station. Where high velocity winds are common strategically located baffles and finger dikes decrease back mixing and moderate wave fetch.

Dunaliella salina is a type of halophile green micro-algae especially found in sea salt fields. Known for its anti-oxidant activity because of its ability to create large amount of carotenoids, it is used in cosmetics and dietary supplements. Few organisms can survive in such highly saline conditions as salt evaporation ponds. To survive, these organisms have high concentrations of β-carotene to protect against the intense light, and high concentrations of glycerol to provide protection against osmotic pressure. This offers an opportunity for commercial biological production of these substances.

From a first pilot plant for Dunaliella cultivation for β-carotene production established in the USSR in 1966, the commercial cultivation of Dunaliella for the production of β-carotene throughout the world is now one of the success stories of halophile biotechnology. Different technologies are used, from low-tech extensive cultivation in lagoons to intensive cultivation at high cell densities under carefully controlled conditions. Although Dunaliella salina produce β-carotene in a high salt environment, archaea such as Halobacterium, not Dunaliella, are responsible for the red and pink coloring of salt lakes. Occasionally, orange patches of Dunaliella colonies will crop up.

Attempts have been made to exploit the high concentrations of glycerol accumulated by Dunaliella as the basis for the commercial production of this compound. Although technically the production of glycerol from Dunaliella was shown to be possible, economic feasibility is low and no biotechnological operation presently exists that exploits the alga for glycerol production.

Nutrient Source, Effect, Control:

Intake water from mangrove thickets, adjacent to upwelling, located near outflow of sewage effluent, or near runoff from fertilized farmland is often rich in nutrients (microorganisms, dissolved and particulate organic substance, ammonium nitrate, and phosphate). Other source of nutrients include bird droppings, floor communities and sediments released by dike destruction, high velocity winds and deep lagoons within the circuit of ponds.

Excessive concentrations of nutrients in a saltwork during prolonged period of time result in production of organic substances greater than consumption of these substances. The resultant accumulations of black organic substances on pond floors and corners decrease surface areas and volumes, deposition of gypsum as microscopic crystals, and delivery of organic-rich brine to crystallizers. Consequences include soft crops and salt floor unable to support machinery and decreased quality and quantity of salt.

Technology for Crystallizing ponds

Crystallizer floors of seasonal and continuously operated Salinas accumulate organic substances, magnesium and other contaminates with contribute to un-desirable crystal characteristics, result in small and hollow layered hoppers, and soft crops and salt floors. Crystallizer floor of seasonal salt works rinsed twice with seawater after harvest, loose much of their accumulated contaminants; floors of continuously operated salt works, when removed and replaced on a routine schedule, also prevent of organic and inorganic contaminants and their consequences.

Artemia : For Nutrient Control and Crystallizer Performance

Artemia grow and reproduce in water of intermediate salinity but many persist in high salinity where they ingest organic particulates, microalgae, gypsum crystals and activities which clear the water. After oxidizing the ingested organic substances filtered from the water, the brine shrimp discard all their wastes in small bags-fecal pellets-which become locked away in the floor sediments. Populations of artemia contribute to the delivery of nutrient-depleted high quality brine to the crystallizers. An additional benefit from Artemia occurs after they reach crystallizers where they disintegrate, feed the resident red halophic bacteria which further decrease organic substances in the brine, increase the red color or the brine, and increase evaporation. Conditions to create and maintain useful populations of Artemia include establishment of appropriate depth and flows in their habitat to control temperatures and management of the upstream ponds to provide suitable food for the brine shrimp.

Fig 9.3: Artemia in condensors

Artemia's feeding and waste disposal activities were utilized in a salina that became dominated by Aphanothece halophytica and its mucilage after hurricane destroyed the floor communities and released massive amounts of nutrients. Large quantities of well feed Artimia continuously introduced into intermediate salinity ponds removed almost all the Aphanothece Halophytica and stopped mucilage production. After reaching crystallizers, the disintegrated Artemia allowed red halophytic bacteria to attain large population, removed most of the remaining mucilage in the brine, and produce bright red colours which aided brine evaporation and high quality salt.

For saltworks management of Artemia we have to process 1. Collection of cyst, 2. Washing & storing of cysts. 3. Hatching and transferring of Nauplius to condensers.

1. Collection of Cysts.

   In general cysts observed in condensers between 15°Bé to 24°Bé. Care full observation require for cyst collection. Cysts observed in the corner of condensers with wave and wind. Observed in floating pink red colored with other impurities like floating debris and feathers. Such cysts must collect with nylon bolting cloth scoop net. In general all the salt works having in different amount artemia and cysts. The migratory birds are the main source of artemia migration. Cyst can spread with birds and migrate in different area. This is the reason Artemia franciscana and Artemia Salina observed to gather.

2. Washing and Storage of cysts.

   After collection wash cysts with sweet water. This process should be quick while washing with sweet water or low density washing, because it may possible to hatching while washing of cysts and will lost all the cysts. There must 2 quick wash first 3.5°Bé wash and then sweet water (0°Bé) wash. After such primary wash pour all the cysts in 11°Bé brine. In this time the cysts will float and unwanted material and soil particle will settle down to bottom. And floating cyst can collect. In sweet water if the cysts floats then these are not fertile cysts and need further time so we have to store in 24°Bé. In 24°Bé we can store long time as long we have to store. The good cysts will settle in bottom of sweet water and will float in 24°Bé Brine. At the storage in brine we have to stair time to time for mixing. Storage in 24°Bé will also increase the quality of cysts also.

Before hatching should preserve cysts in 24°Bé at least for 10 days for good result of hatching. If we do not want to use the cysts then we can preserved washed, dried & screened cysts for long time. For preservation it should pack in air tight packing.

3. Hatching and transferring of Nauplli to condensers.

Hatching needs care full and controlled environment. 100 to 200 gram of cysts can put in hatching tank with capacity of 200 liter. Tank should be filled with filtered clean 3.5°Bé brine. For hatching temperature should maintain up-to 32°C and continuous aeration. In 12 to 24 hours cysts will hatch and nauplii observes in hatching tank. Eggs hatch into nauplii that are about 0.5 mm length. They have one single simple eye that only senses the presence and direction of light. Nauplii swim towards the light but adult individuals swim away from it. For collection remove aeration and let nauplii to sattle at the bottom of the tank. For better result can put light source at the bottom of the tank. Such collected nauplii can release to condensors in concentration higher than 15 °Bé brine.

Fig 9.4: Artemia Hatchery

Biological Systems – Favorable & Unfavorable

The Physical phenomena of evaporation and precipitation of low and high solubility salts are intimately linked to biological processes that occur in every pond of solar saltworks. Because brine organisms and the communities they produce are essential to salt production, knowledge of the structure, function, and management of the biological system is important for saltworks operations.

The Brine Biological System

Attached seaweeds and sea grasses, small animals, and microscopic organisms inevitably develop a biological system composed of planktonic and benthic communities in the ponds of every solar salt works. The plants, algae and bacteria with photosynthetic pigments use sunlight energy and inorganic nutrients to manufacture organic matter, but the entire group of organisms consumes and oxidizes these substances. The biological system can help or harm salt production.

Biological system that help salt production: develop and maintain planktonic communities consisting of many well represented species that are adapted to narrow salinity ranges, from mat like benthic communities of many well represented species with appropriate quantities of organic matter on pond floors; these are defined as mat layers of sufficient thickness and composition to prevent leakage and sequester nutrients from the overlaying water, minimize accretion of organic substance and gypsum on pond floors, and produce crops of salt in crystallizers able to support heavy harvest machinery. The biological systems are associated with saltworks that continuously and economically produce high quality salt at design capacity.

Biological systems that harm salt production: Produce communities composed of few species that thrive in a wide range of salinities, from inadequate or excessively developed benthic communities that release damaging quantities of mucilage and result in low quality salt at less than design capacity. These biological systems necessitate increased effort and expanse to bring the quality of the harvested product to world standards.

Concentration of dissolved solids, kinds of organisms and physical phenomena that occur from seawater salinity at the intake to brine saturated with sodium chloride in the crystallizers; all divide a solar saltworks into low, intermediate, and high salinity ponds. Each group of ponds has a high salinity ponds. Each of ponds has highly characteristics biological system. As planktonic organisms adapted to particular salinity range flow downstream and reach more saline conditions they die, release their contents, and provide nutrients for the biota in the saltier locations. The succession terminates with the planktonic aerobic red halophilic bacteria in the highly saline ponds and crystallizers.

Because insufficiently developed communities typified by clear water, lack of mats on pond floors, and brine leakage are relatively rare, this condition will not be considered here. Insufficiently developed systems, the consequence of nutrient poor intake water, can be corrected by appropriate supplements of fertilizers to the pond.

Characteristics of Low Salinity Ponds with Favorable and Unfavorable Biological Systems. (Bé 3.5° to Bé 9° : SG 1.025 to SG 1.067)

Favorable Biological Systems:

Of the ponds in the saltworks, those of low salinity occupy the most surface area and volume, contain the largest variety of organisms, and produce the most biomass. More biomass is manufactured than consumed in these ponds. The planktonic community (algae, bacteria, and protozoa suspended in the water) provides organic nutrients for the entire saltworks, colors the water, aids evaporation, and allows light to reach pond floors. The benthic community (mats on the floors of the ponds) consists of an upper most layer of microorganisms similar to those of the plankton as well as small and microscopic animal life like crustaceans, molluscs, nematodes, ostracods, protozoa; a second layer of bacteria, and black organic matter. Benthic communities maintain a diverse biota, and they remove and lock away important quantities of organic substances, combined nitrogen and phosphate from the over laying water.

The mat also moderates leakage through pond floors and maintains desired thickness. Firm adherence of the benthic community to the floors is aided by patches of rooted sea grasses and attached seaweeds. During strong winds, part of the uppermost layer of the mat become suspended and imparts additional nutrients and color to the water. Exports from these ponds (organisms, inorganic and organic nutrients) fuel the biota of the downstream systems at desired levels.

Unfavorable Biological System:

These result from excessive nutrients in the intake seawater, depths and turbidity of brine that intake seawater, depths and turbidity of the brine that prevent light from reaching the benthic community and disturbance (high velocity wind, severe wave action, back mixing of brine, brine dilution, breach of dikes). Unfavorable systems develop large populations of plankton composed of a few species which are able to tolerate the entire salinity range. These suspended organisms shade out the sea grasses, the seaweeds, and the benthic community. When the condition remains uncorrected, sections of benthic communities detach from the floors, partially dissolve, and flow downstream. Export of excessive quantities of microorganisms inorganic and organic nutrients from planktonic and damaged benthic communities create unfavorable systems in the downstream ponds.

Characteristics of Intermediate Salinity Ponds with Favorable and Unfavorable Biological Systems. (Bé 10° to Bé 19° : SG 1.075 to SG 1.152)

Favorable Biological Systems:

Volume of water, surface area, and variety of organisms are smaller than in the low salinity ponds. More organic matter is consumed than is produced in these ponds. Imports from the upstream ponds, and microorganisms produced within the intermediate salinity ponds are maintained at desired levels by the resident suspended bacteria and by grazing brine shrimp (Artemia), chironomid and brine fly larvae. Appropriate concentrations of nutrients, competition for resources among the many species, and stable salinity gradients throughout the ponds restrain undesirable algae from becoming dominant. The essentiality of Artemia (brine shrimp) to salt production lies in the ability of the animals to clear brine of particles up to 50 micrometers dia. , to metabolize large amounts of ingested organic matter to carbon dioxide, to deposit wastes in fecal pellets that become incorporated in the benthic community, and to furnish highly suitable food for the Halobacterium salinarium populations in the downstream ponds.

Benthic communities seal ponds against leakage and maintain a diverse biota of well represented organisms in their uppermost layers. Their lowermost layers (bacteria and black organic substances) remain firmly attached to pond floors, maintain desired thickness, and sequester important quantities of organic matter, combined nitrogen, and phosphate from the overlying water. Organisms in the mat are adapted to narrow segments of the salinity range in which they occur.

Unfavorable Biological System:

Excessive nutrients imported from upstream, disturbances, and inappropriate management of brine depths and flows result in massive reproduction of organic-releasing algae in the plankton, release of damaging quantities of organic substances and mucilage to the brine, decreased levels of dissolved oxygen, and decreased biodiversity. Complexation of calcium ions with high levels of organic matter causes calcium carbonate to precipitate at higher than expected salinities, brine turbid with carbonates and gypsum, and decreased evaporation. These phenomena create viscous brine and environments highly unfavorable to Artemia, resulting in increased calcium and sulfate concentrations in the harvested salt.

Excessively developed systems are caused by high concentrations of nutrients in the intake water, droppings from large numbers of waterfowl, accumulated organic substances in the ponds, inappropriate saltworks design, insufficient competition among organisms for nutrients (due to dominance of one or more species), fluctuating salinities within ponds, and disturbances. Excessively developed benthic communities dominated by organic-releasing algae become gelatinous, trap gases, accrete excessively, and decrease pond surface areas and volumes. If the condition is not corrected, patches of the community detach from the pond floor, float to the surface, partially dissolve, increase the viscosity of the brine, decimate the Artemia population, and contribute damaging quantities of organic substances to all the downstream ponds.

Characteristics of High Salinity Ponds with Favorable and Unfavorable Biological Systems. (Bé 20° to Bé 25° : SG 1.161 to SG 1.210)

Favorable Biological Systems:

Of the concentrating circuit, the high salinity ponds have the least surface area and volume, and the smallest variety of organisms. Their planktonic communities are dominated by Halobacterium salinarium (aerobic red halophilic bacteria); Dunaliella salina is present in low concentrations. Nutrients from upstream (mainly living and dead organisms, particulate and dissolved organic matter) power the biota; little to no organic matter is manufactured by photosynthetic organisms in these ponds.

When appropriate quantities of nutrients (These are defined as the quantities and quality of organic substances in crystallizers that enable Halobacterium salinarium to maintain their population at several billions per liter, and that keep concentrations of Dunaliella salina low) reach the ponds, the planktonic Halobacterium salinarium color the brine pink to bright red-hues that aid solar energy absorption and brine evaporation, and through their rapid metabolism maintain organic matter at desired levels. The firm gypsum deposit (calcium sulfate) remains attached to pond floors, accretes only slightly over time, is free of adherent mucilage, and covers large areas of the bottoms. A three layered benthic community blue-green algae uppermost, purple bacteria (Chromatium and Thiocapsa) in the center, and black organic matter lowermost, able to remove and lock away organic and inorganic sub stances exists within the gypsum deposit, under the gypsum, and on floor areas without gypsum. Bacteria associated with the deposit erode the lowermost part of the gypsum to a fine sand part of which may be volatilized. These activities decrease the accretion rate of the calcium sulfate and black substances on pond floors. Brine delivered to crystallizers is nearly free of particulate gypsum.

Unfavorable Biological System:

Excessive nutrients from the upstream ponds, excessive quantities of imported planktonic Aphanothece halophytica and mucilage from upstream, development of large Dunaliella salina populations, and accumulations of Aphanothece halophytica at the gypsum-water interface result in viscous brine, quantities of suspended organic substances the aerobic halobacteria are unable to control, and fast accretion of sediments and gypsum on pond floors and peripheries. Complexation of calcium salts with the organic matter and precipitation of gypsum as widely separated microscopic crystals that stay suspended in viscous brine or that become loosely embedded in mucilage on pond floors result in the downstream flow of excessive quantities of dissolved and particulate calcium sulfate to the crystallizers.

Characteristics of Crystallizer Ponds with Favorable and Unfavorable Biological Systems. (Bé 25.5° to Bé 30° : SG 1.216 to SG 1.264)

Favorable Biological Systems:

Appropriate nutrient input from upstream (e.g., sufficient quantities of dead brine shrimp, dissolved and particulate organic substances) allows the resident Halobacterium salinarium population to color the water and maintain the organic content in the brine at desired levels. Under these conditions the Dunaliella salina population remains small and contributes insignificant quantities of organic matter to the brine. The salt deposit (crop), crystallizer perimeters, and pond floors remain firm and well able to support heavy equipment. Harvested sodium chloride crystals are mostly solid, and they range from about one mm to more than three centimeters in the longest dimension. The wash process results in losses of about 12 to 15%, and the residence time of the salt crystals on the stockpile is comparatively short in order to bring the harvested product to world standards.

Unfavorable Biological System:

Excessive quantities of organic matter imported from upstream as well as organic release from large populations of Dunaliella salina living in the crystallizers may exceed the ability of Halobacterium salinarium to prevent the substances and algae from accumulating. Organic-rich brine may become viscous, blackened, and depleted of dissolved oxygen, and it decimates the Halobacterium salinarium population. Because organic substances interfere with normal crystal formation, the salt deposit in crystallizers becomes soft with fine particle size and may not support heavy harvest equipment. Soft salt first appears at crystallizer perimeters, but the width of the soft salt may increase to include the entire deposit. Crystals in the harvested crop are mostly layered, hollow, hopper-shaped pyramids; they retain bacteria; and they trap liquid and crystalline contaminants between layers, in cavities, and on their surface. The weight of trapped liquid within salt crystals increases fuel consumption in harvest equipment, and drainage from vehicles that haul brine-laden salt damages roads and dikes. Furthermore, small and hollow crystals decrease the efficiency of the wash process, and they may require complex wash equipment, dual washing, use of caustic compounds, and long residence on the stockpile to prepare the product for market.

Practical Management of the Biological System:

Except for the few salinas whose design and location complement the physical and biological processes in the ponds to achieve economical and continuous production of high quality salt at design capacity, the biological system in all other saltworks must be managed to achieve these results. Management requires continuously gathered data and appropriately displayed information on the status of the biological and physical systems in each set of ponds, as well as constant vigilance and upkeep of the pond system and associated equipment.

Continuous maintenance of the salinity gradient within each pond and between ponds in narrow and unchanging limits during the production season sustains a maximum diversity of organisms suited to each salinity range. Implementation to keep salinity constant at any place in the concentrating ponds can be accomplished by appropriately adjusting flow rates to match evaporation, by placing strategically located dikes that partially or completely traverse ponds to prevent back-mixing and decrease wave fetch, by providing standby pumps at each pump station, by preventing land runoff from entering ponds, and by removing rainwater with decanters.

Management of Key Organisms

Artemia

By converting much of the massive plankton imported from upstream into their own substance, Artemia in the intermediate salinity ponds provide an important link between the biota of the low and high salinity ponds (Davis, 1990). In many saltworks, the local Artemia strain is self maintaining, sufficiently aggressive to clear the brine, and able to quickly develop functioning populations after the desired salinity gradient has been reestablished. However, re-introduction of Artemia may be required when numbers of the animals are too small to control the plankton and suspended organic particulates. New introductions are indicated when the population of the local strain remains too small or performs poorly. Maintenance of Artemia populations able to accomplish activities essential to salt production can be aided by increasing brine depths in the intermediate salinity ponds for greater production of suitable food and by preventing brine temperatures from exceeding 35 °C, by maintaining the salinity within and between ponds in a narrow and constant range, and by modifying ponds to minimize large accumulations of the animals and cysts in windward corners. Implementation of management practices that allow favorable systems to develop in the low salinity ponds help to provide suitable food and appropriate conditions for brine shrimp in the intermediate and high salinity ponds.

Aphanothece halophytica

These euryhaline, slime-producing algae grow best in intermediate salinities, but the organisms survive and continue to release mucilage well into high salinities. Techniques that minimize growth of Aphanothece halophytica include establishment of sufficient numbers of ponds to keep the salinity gradient in each pond narrow, adjustment of depths in low and intermediate salinity ponds to allow light to reach the pond floors and permit nutrient-sequestering benthic communities to develop and function.

Dunaliella Salina

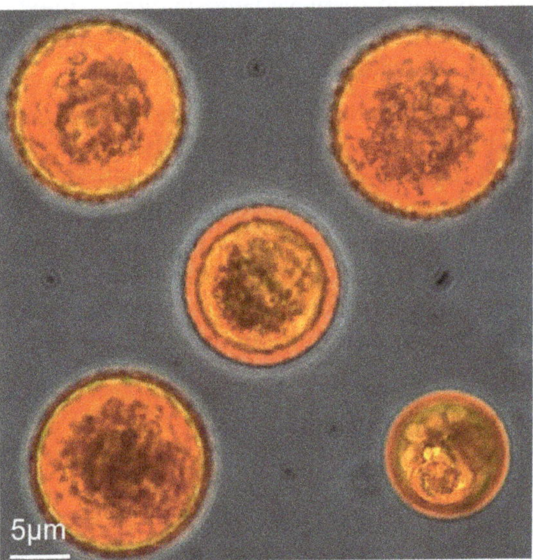

Fig 9.5: Dunaliella Salina

Although *Dunaliella Salina* reproduce best in intermediate salinities, the organisms continue to divide slowly in the high salinity and crystallizer ponds. When excessively high concentrations of nutrients reach the high salinity and crystallizer ponds, Dunaliella Salina may form populations sufficient to color the brine red-orange and release damaging quantities of organic substances. Concentrations of Dunaliella Salina can be kept low by management practices that aid maintenance of large Artemia populations in the intermediate salinity ponds and functioning nutrient sequestering benthic communities and stable salinities in the low and intermediate salinity ponds.

Halobacterium salinarium

Management techniques to obtain high concentrations of these bacteria in the high salinity ponds and crystallizers begin by maintaining favorable biological systems in the low and intermediate salinity ponds. In the highly saline ponds, shallow brine depths and long residence of brine allow sufficient concentrations of oxygen to diffuse into the brine to maintain high concentrations of *Halobacterium Salinarium.* Because petroleum products are toxic to the bacteria, care must be taken to prevent machinery used in crystallizers from leaking fuel, grease, or hydraulic fluids. Effects of spills can be minimized by quickly removing leaking vehicles and by extracting salt and soil contaminated with petroleum products.

10
Mixed Salt

Introduction

Potassium fertilizers are essential as plant nutrients and are required in large quantities. Also several to other potassium salts are used in commerce. Most of the world's demand is meet from bedded deposits of soluble salts chiefly as chloride and sulphate. In India, there are no potassium deposits and requirement of the country is entirely met by imports. Sea-water and certain inland sub-soil brine contain potassium and are the main source of potassium chemicals in the country.

Manufacture of salt from sea-water and inland salt brine (Sub soil brine) in this country is exclusively by solar evaporation. After the salt separation between 24° and 29°Bé, the solution of 29-30°Bé left over is called bittern and contains 14 to 15 gram of NaCl, 10 to 12 gram of $MgCl_2$, 6 to 8 gram of $MgSO_4$ and 2 to 2.5 gram of KCl per 100 ml of solution. As such, it is a poor source and hence processes based on bittern should be well controlled to be efficient or, alternatively, bittern should be evaporated further to obtain potassium in more concentrated and easily extractable form. Mixed salt, a material much richer in potassium chloride than bittern is collected by solar evaporation of bittern. The separation of mixed salt from bittern can be achieved at lower densities with addition of required quantity of magnesium chloride in the form of 36°Bé bittern produced at the end of evaporation cycle thereby saving the time required for evaporation. The evaporation of mixed bittern is carefully controlled and carried out in two steps; first to 34 to 34.5°Bé, to separate sodium chloride as crude salt and then to 36-36.5°Bé to precipitate mixed salt containing 18 – 20 percent potassium chloride and about equal concentration of sodium chloride. The concentration of magnesium sulphate in inland and sea bittern varies; therefore, mixed salt from sea bittern contains 30 – 35 percent magnesium sulphate, while that from inland brine contains lower proportion of magnesium sulphate depending on the source of bittern. Magnesium chloride is only from adhering mother liquor and varies between 6 to 8 percent.

The bittern is further concentrated by solar evaporation to reach 36°Bé to 36.5°Bé; the composition of 36.5 °Bé bittern is approximately: $MgCl_2$, 34.7; $MgSO_4$ 7.1; KCl 1.5 and NaCl 2.5 percent. Seventy percent of the total potassium chloride present in the original 29-30°Bé bittern separates as mixed salt. The final checking of the bittern at 36°Bé is also essential. The normal composition of mixed salt is: $MgSO_4$ 30-35%, $MgCl_2$ 6-8%, NaCl 15-20% and KCl 18-20% (wet analysis). Yield of mixed salt per 454.6 liter of bittern is approximately 39 – 40 Kg.

At the beginning of the season when 36°Bé bittern is not available for addition, direct evaporation of sea bittern from 30°Bé to 38°Bé is to be carried out only during the first month. This is also carried out in two steps; first from 30 to 35.5°Bé where in most of the sodium chloride separates as crude salt. Bittern of 35.5-36°Bé is ready for mixed salt crystallization only when **concentration of both sodium chloride and potassium chloride are nearly equal**. At this stage it is led to mixed salt crystallizing pans and allowed to evaporate to 38°Bé when most of the potassium chloride separates out as mixed salt. The concentration of potassium chloride in the end liquor (38°Bé) should be 1.5 to 1 %. After one or two direct evaporation cycles it is preferable to recycle the bittern as described above. This process also depends on atmospheric conditions like humidity in atmosphere, temperature, wind-speed and wind direction. Content of magnesium in bittern pulls moisture from atmosphere, which leads this process slower than assumed.

Preparation of Mixed Salt Pans

Location

The land selected for layout of mixed saltpans should be nearer to salt crystallizing pan area, where all the bulk of the discharged bittern is available, but should not be in very closer proximity so as to contaminate and hinder the production of good quality.

Area requirement

Area requirement for mixed salt production, has been arrived at after field trials. It depends upon various factors like climatic conditions of the place, period available for bittern evaporation and methods of discharge of bittern. If bittern supply is continuous right from the month of November, even 2.20 hectares of land is sufficient for the production of 1000 ton of mixed salt. This does not include area for storage platforms, channel and reservoirs for **bittern and crude salt ponds**.

Preparation of bed

The mixed salt beds are prepared similar to salt crystallizing pans. The bed are carefully prepared to minimize the percolation losses and are made sufficiently hard so that mixed salt is not contaminated with fine clay. To get good quality of mixed salt we can use HDPE lining to eliminate percolation losses. This is one time expanse but the result is good in term of expanse.

Layout of Pan

The bulk of the discharged 29-30°Bé bittern is led first to bittern reservoir; this is sufficiently large to accommodate the discharge from a single set of pans. A pumping set for lifting of bittern is necessary at this point. The height of the bittern thus raised helps further conveying to mixed salt pans by gravity alone, or need additional pump set to feed mixed salt pans. The mixed salt pans may be of any suitable size. The layout of pans is changed from place to place so as to suit the needs of the salt farm. However the division of pans in to condensers for crude salt deposition and mixed salt crystallization remains practically constant as it depends on the reduction of volume of the bittern during various stages of evaporation. The volume of 30°Bé bittern is practically reduced to one half and one third while its concentration increase to 34.5°Bé and 36.5°Bé the time it enters and leaves the mixed salt area. On this basis for every three heactares of land for mixed salt production, two heactares are used for condensers area and one heactare for mixed salt crystallizing pans.

Addition of dye:

Evaporation of bittern after 34°Bé is very slow. In order to facilitation of evaporation and obtain the mixed salt in short period, the bittern can be treated with solivap green at the rate of 11.2 kg per hectare. The dosage varies with the depth of bittern in the salt pans. The laboratory data shows that with the addition of magnesium chloride and solivap green, the period of evaporation is reduced from 20 days to 9 days to reach 36°Bé.

Operation control in mixed salt production.

The following two methods can be used for controlling the crystallization of mixed salt and to obtain the product containing 18-20 percent potassium chloride. (1) Density control, (2) Analytical control.

1. Density Control:

   Mixed salt is produced by evaporation of sea bittern of 29°Bé first upto 36°Bé whereby crude salt is separated and then further evaporated up to 38.0°Bé whereby mixed salt containing 18% to 20% KCl is obtained. It has been also noticed that by addition of 36°Bé bittern to 29°Bé bittern, in 1:1 proportion, mixed salt can be crystallised earlier i.e. between 34.5°Bé to 36.5°Bé. When bittern of 36°Bé and 29°Bé are mixed in the above mentioned proportion resulting density becomes 30 to 31°Bé. This mixed bittern is further subjected to evaporation upto 34°Bé to 34.5°Bé and crude salt deposited is collected separately. Further 34.5°Bé bittern is charged into mixed salt crystallizing pans and evaporated up to 36.5°Bé which gives mixed salt of desired composition. Thus, by determining the density one can produce the mixed salt but as the temperature affects the density, the density could be determined at fixed temperature say at 28°C and the sample should be filtered to obtain the correct density.

2. Analytical control:

   For very precise control one can follow this method. The control can be based upon potassium chloride and sodium chloride analysis at various stages during evaporation of bittern using flame photometer. Potassium can be also determined by gravimetric method using reagent sodium tetraphenyl boron.

   Potassium chloride and Sodium Chloride control:

   Potassium and sodium ions can be quickly determined by flame photometric method. Salt manufactures having better laboratory facilities can adopt this method for mixed salt crystallization control. There exists a particular range of concentration of potassium chloride and sodium chloride at states, before at and after mixed salt crystallization. The bittern of 29°Bé contains about 2.0 gram per 100 ml of potassium chloride and 18.5 gram of sodium chloride. In mixed bittern at 34 to 34.5°Bé, concentration of both these salts become more or less similar. At this stage bittern is saturated with potassium chloride and ready for the mixed salt crystallization. Again at 36.5°Bé when most of the potassium chloride is thrown out of the system the concentration of each become 1 to 1.5 gram (KCl) and 2 to 2.5 gram (NaCl) per 100 ml respectively. At this stage bittern is discharged and fresh charge of 34.5°Bé bittern is let in. Thus by determining potassium and sodium ion one can control the operation of crystallization and obtain the mixed salt of desired quality.

Stage wise bittern analysis

Density	°Bé	30.1	31.6	34.0	34.5	36.0	36.5	37.5
Volume	%	100	87	68	61	47	38	28
KCl	gr/100ml	2.50	2.54	3.50	3.85	4.15	2.23	0.90
NaCl	gr/100ml	13.00	9.50	6.50	4.50	3.00	2.30	1.15
$MgCl_2$	gr/100ml	13.90	17.06	19.17	21.88	28.35	33.10	36.9
$MgSO_4$	gr/100ml	8.65	9.77	12.20	10.78	8.02	6.17	6.10

Evaporation of mixed bittern (Ratio of addition 1:1) (Composition % w/v)

Mixing of 30.1°Bé bittern with 36.1°Bé equally.

Density	30.1	36.1	33.5	34	34.5	35.5	36.5
volume %	50	50	93	82	76	67	56
Composition	% w/v	% w/v	% w/v	% w/v	% w/v	% w/v	% w/v
KCl	2.50	0.70	1.68	1.75	1.87	0.60	0.38
NaCl	13.00	1.15	7.57	2.70	1.63	0.90	0.50
$MgCl_2$	13.90	36.94	26.88	29.43	33.00	37.00	40.60
$MgSO_4$	8.65	6.10	7.00	8.30	7.10	6.15	5.50

Salt fractions (Evaporation of mixed bittern 1:1)

Stage	Immediately at 31° Bé	32 to 33.5°Bé	33.5 to 34.5 °Bé	34.5 to 35.2 °Bé	35.2 to 36.5 °Bé
KCl	0.62	2.55	1.25	19.40	17.9
NaCl	79.5	69.2	22.5	14.74	15.9
$MgCl_2$	4.43	4.23	1.51	6.95	11.95
$MgSO_4$	4.66	9.92	35.3	28.4	26.5

Storage of Mixed Salt:

All the four constituents of mixed salt are reactive and also corrosive. Main reaction are taking place between chlorides of potassium and sodium on the one hand and magnesium sulphate on the other. The product of reaction is usually magnesium chloride and a less soluble salt. Water is essential for the reactions and the rate and direction of the reaction is controlled by concentration of the reaction product and how quickly it is drained off. In case of rain and intermittent shower the reactions are fast enough and go to completion with in few days. In case of very humid atmosphere prevails either due to wave progress but slowly. Mixed salt as a rule, therefore, cannot be store in open space. It should be store under shade and should be protected from shower or water sprays of any kind even during transportation.

In case of carnallite type mixed salt containing a mixture of sodium chloride and carnallite with very much less or no magnesium sulphate, storage difficulties are much more enhanced. Such mixed salt cannot be store in open shade. Preferably, the carnallite type mixed salt should be decomposed by water or low density brine and the resulting unreactive mixture of sodium and potassium chlorides can then be stored in open shades.

Completely enclosed structure is ideal but a costly proposition. Open shade with corrugated cement or galvanized iron sheets or prefabricated structure are suitable. As the mixed salt is corrosive, aluminum sheets are unsuitable. Wooden structure or mild steel structure with anticorrosive paints is cheap and suitable.

Some bittern and solid analysis for Mixed Salt.

Liquid and Solid analysis in mixed salt crystallizer

Liquid Samples	
Parameters	
Sp.gr.(g/cc)	1.3353
° Bé	36.41
	gpl
Ca	0
Mg	99.02
Cl	265.95
SO_4	61.99
Bromine	3.52

Solid Samples	
Parameters	**% As such**
W.I.	8.67
Ca	0.098
Mg	6.63
Cl	22.15
SO_4	20.49
Bromine	0.09
KCl	16.93
Compositions	

Liquid Samples	
Parameters	
KCl	23.08
Compositions	
Parameters	gpl
$CaSO_4$	0.00
$MgSO_4$	77.69
$MgBr_2$	4.06
$MgCl_2$	324.45
KCl	23.08
NaCl	21.91

Solid Samples	
Parameters	**% As such**
Parameters	% as such
$CaSO_4$	0.33
$MgSO_4$	25.39
$MgBr_2$	0.10
$MgCl_2$	5.82
KCl	16.93
NaCl	16.08
W.I.	8.67
Moist.(diff.)	26.68
TOTAL	100.00

Solid samples collected before harvesting

Solid Samples	**M1**	**M2**
Parameters	%	%
	As such	As such
W.I.	2.30	0.87
Ca	0.03	0.00
Mg	7.64	8.26
Cl	20.99	19.15
SO_4	25.18	28.68
Bromine	0.12	0.07
KCl	20.19	22.78

Compositions	**M1**	**M2**
Parameters	% as such	% as such
$CaSO_4$	0.10	0.00
$MgSO_4$	31.47	35.95
$MgBr_2$	0.14	0.08
$MgCl_2$	4.94	3.86
KCl	20.19	22.78
NaCl	12.68	8.95
W.I.	2.30	0.87
Moist.(diff.)	28.18	27.52
TOTAL	100.00	100.00

References

- Tata Chemicals Ltd. IISS, 2010.
- Research and Industry, 37, 1992, 46.
- CSMCRI, Bhavnagar, IISS, 2010.
- Guy Wilkins, IISS, 2010
- Biology, The Neglected technology for the manufacture of solar salt. Joseph S Davis
- B. S. Joshi & R. B. Bhatt CSMCRI, SISS, 1983
- Factor determining the rate of solar evaporation in the production of salt. : C. Warren Bonython. 1965
- Sodium Chloride: Dale W. Caufmann
- Salt Matters, March 2002 Vol:1
- Structure, Function, and Management of the biological system for seasonal solar saltworks: Joseph S Devis: 2000
- Salt: Technology & Manufacture of By-Product by Kapilram H. Vakil

- Internet
- www.tatasalt.com
- www.wikipedia.org
- Google Earth

Appendices

Appendix 1: Separation of Salt on Different Density

Density ° Bé	Salts
3.5° Bé	Sea Water
6.0° Bé	Insoluble & Fe_2O_3
8.0° to 14° Bé	$CaCO_3$
14° to 17.5° Bé	Traces of $CaCO_3$ & $CaSO_4 \cdot 2H_2O$
17.5° to 24.5° Bé	$CaSO_4 \cdot 2H_2O$ (Gypsum)
24.5° to 27.5° Bé	$CaSO_4 \cdot 2H_2O$, NaCl, $MgCl_2$ & $MgSO_4$
27.5° to 29.0° Bé	NaCl, $MgCl_2$ & $MgSO_4$
29.0° Bé	(Bittern for Bromine recovery)
29.0° to 33.5° Be	NaCl, $MgSO_4 \cdot 7H_2O$ (Crude Salt)
33.5° to 36.0° Bé	NaCl, $MgSO_4 \cdot 7H_2O$ (Sales Mix Salt)
36.0° to 38.0° Bé	Mix Salt KCl, $MgSO_4 \cdot 7H_2O$, $MgCl_2 \cdot 6H_2O$ (Carnallite)
38.0° Bé to solidification	$MgCl_2 \cdot 6H_2O$ (Carnallite)

Appendix 2: Composition of dissolved salts in brine during progressive evaporation of sea water Grams of Salts per liter of brine as a function of brine density g/liter of brine at 22.2° C

° Bé	Sp.Gr.	Dissolved Salts, in gms. per liter (gpl)						Total Salts	H_2O
		$CaSO_4$	$MgSO_4$	$MgCl_2$	NaCl	KCl	NaBr		
0.72	1.005	0.340	0.493	0.799	6.38	0.173	0.012	8.20	996.80
1.44	1.010	0.614	0.900	1.457	11.62	0.315	0.037	14.94	995.06
2.14	1.015	0.884	1.320	2.107	16.92	0.460	0.053	21.75	993.25
2.84	1.020	1.160	1.738	2.768	22.20	0.605	0.070	28.55	991.45
3.50	1.025	1.419	2.138	3.384	27.27	0.711	0.085	36.04	989.66
4.22	1.030	1.711	2.598	4.076	33.01	0.897	0.104	42.41	987.59
5.58	1.040	2.265	3.468	5.403	43.91	1.192	0.138	56.29	983.61
6.91	1.050	2.816	4.367	6.719	54.98	1.491	0.172	70.56	979.44
8.21	1.060	3.377	5.277	8.045	66.19	1.795	0.208	84.91	975.09
9.49	1.070	3.933	6.209	9.374	77.53	2.101	0.242	99.41	970.59
10.74	1.080	4.499	7.157	10.710	89.02	2.414	0.278	114.10	965.90
11.94	1.090	5.048	8.096	12.040	99.92	2.709	0.313	128.33	961.47
11.97	1.089	5.046	8.110	12.420	100.23	2.717	0.314	128.56	961.44
13.18	1.100	4.695	9.197	13.600	113.08	3.062	0.354	144.80	956.00
14.37	1.110	4.360	10.290	15.070	125.96	3.408	0.393	159.52	950.48
15.54	1.120	4.011	11.360	16.580	138.85	3.765	0.435	175.05	954.95
16.68	1.130	3.746	12.430	18.050	151.78	4.105	0.473	190.64	939.36
17.81	1.140	3.543	13.510	19.640	164.76	4.456	0.515	208.33	933.67
18.91	1.150	3.171	14.580	21.190	179.09	4.807	0.555	222.06	927.94
20.00	1.160	2.898	15.690	22.740	190.72	5.159	0.596	237.79	922.21
21.07	1.170	2.628	16.740	24.350	203.53	5.518	0.636	253.68	916.32
22.12	1.180	2.369	17.830	25.940	217.11	5.875	0.679	269.90	910.10
23.15	1.190	2.228	18.810	27.640	230.02	6.228	0.720	285.64	904.36
24.17	1.200	1.862	20.030	29.180	248.81	6.588	0.760	302.05	897.95
25.17	1.210	1.606	21.130	30.810	256.81	6.953	0.802	318.17	891.83
26.00	1.219	1.400	22.060	32.280	268.38	7.266	0.840	332.29	886.21
26.15	1.220	1.346	23.780	35.660	263.03	7.921	0.915	332.99	887.01
27.11	1.230	1.064	27.830	54.980	233.54	12.360	1.500	341.40	888.60
28.06	1.240	0.817	50.820	74.050	206.45	16.630	1.910	350.79	889.21
28.53	1.245	0.703	57.110	83.420	193.20	18.710	2.160	356.38	889.62
29.00	1.250	0.595	63.240	92.710	180.34	20.760	2.390	360.15	889.85

Note:

1. Table above indicates the Composition of Salts in Brine at Various stages of Evaporation as a function of Brine Density at 22.2° C.
2. The table gives typical values - but figures may vary slightly due to changes of Sea Water source, etc. and is to be used only as a general guideline and reference.

Appendix 3: Composition of dissolved salts in brine during progressive evaporation of sea water

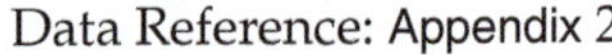
Data Reference: Appendix 2

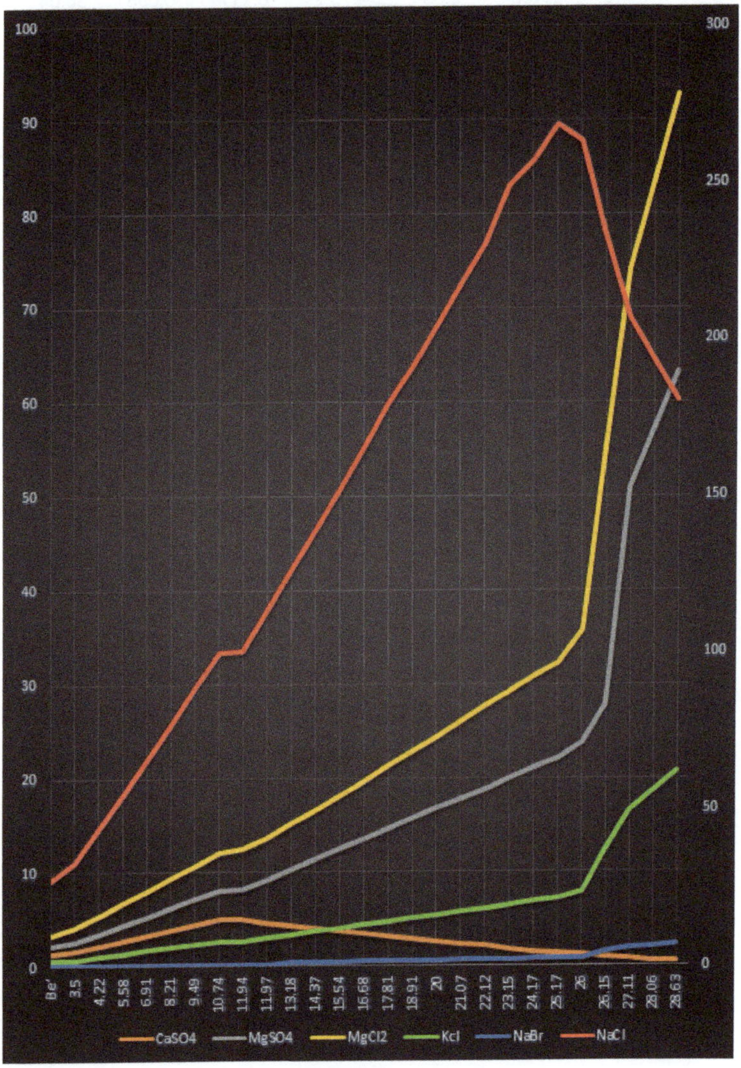

Appendix 4: Density – Evaporation Relationship

Density ° Bé	Evaporation	Density ° Bé	Evaporation	Density ° Bé	Evaporation
0	1	10	0.88	21	0.748
1	0.988	11	0.868	22	0.736
2	0.976	12	0.856	23	0.724
3	0.964	13	0.844	24	0.712
3.5	0.958	14	0.832	25	0.7
4	0.952	15	0.82	26	0.67
5	0.94	16	0.808	27	0.64
6	0.928	17	0.796	28	0.61
7	0.916	18	0.784	29	0.58
8	0.904	19	0.772	30	0.55
9	0.892	20	0.76	31	0.52

Appendix 5: Evaporation -> Density

Data Reference Appendix 4

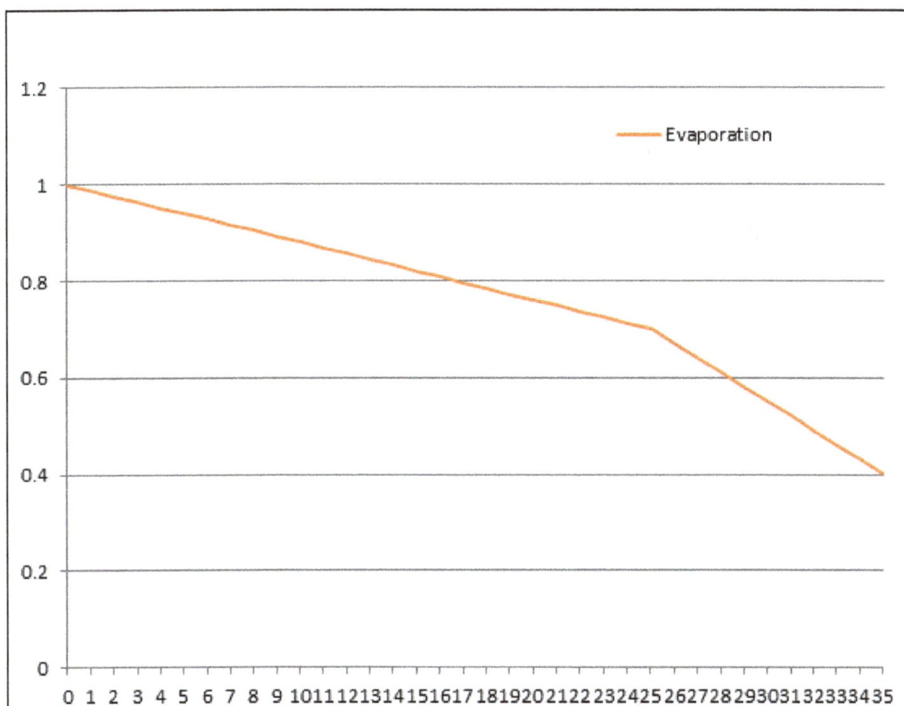

Appendix 6: Elements in Brine

Grams of loans per liter of brine as a function of brine density g/liter of brine at 22.2° C

° Bé	SpGr	Ca	SO_4	Mg	Cl	K	Na	Br	Total Salts	H_2O
0.72	1.005	0.100	0.630	0.304	4.55	0.091	2.51	0.016	8.200	996.800
1.44	1.010	0.181	1.151	0.553	8.29	0.166	4.58	0.028	14.940	995.060
2.14	1.015	0.257	1.677	0.805	12.06	0.242	6.67	0.041	21.750	993.250
2.84	1.020	0.342	2.206	1.056	15.82	0.317	8.75	0.054	28.550	991.450
3.50	1.025	0.418	2.708	1.296	19.42	0.389	10.74	0.067	35.040	989.660
4.22	1.030	0.504	3.278	1.566	23.49	0.471	13.01	0.080	42.410	987.590
5.58	1.040	0.667	4.367	2.080	31.24	0.625	17.31	0.107	56.390	983.610
6.91	1.050	0.830	5.472	2.598	39.08	0.782	21.66	0.133	70.560	979.440
8.21	1.060	0.993	6.593	3.120	47.01	0.941	26.09	0.161	84.910	975.090
9.49	1.070	1.159	7.732	3.649	55.03	1.102	30.55	0.188	99.410	970.590
10.7	1.080	1.324	8.887	4.182	63.15	1.265	35.08	0.216	114.100	965.900
11.9	1.090	1.486	10.020	4.710	70.98	1.421	39.27	0.243	128.230	961.470
12.0	1.090	1.486	10.032	4.735	71.14	1.424	39.50	0.244	128.560	961.440
13.2	1.100	1.382	10.650	5.328	80.20	1.606	44.56	0.275	144.000	956.000
14.4	1.110	1.284	11.290	5.926	89.29	1.787	49.64	0.305	159.520	950.480
15.5	1.120	1.191	11.920	6.527	98.39	1.970	54.71	0.337	175.050	944.950
16.7	1.130	1.102	12.570	7.134	107.50	2.153	59.81	0.368	190.640	939.360
17.8	1.140	1.017	13.210	7.744	116.70	2.337	64.92	0.400	206.330	933.670
18.9	1.150	0.934	13.880	8.358	125.90	2.521	70.04	0.431	222.060	927.940
20.0	1.160	0.853	14.550	8.977	135.10	2.705	75.14	0.463	237.790	922.210
21.1	1.170	0.775	15.220	9.601	144.50	2.893	80.20	0.495	253.680	916.320
22.1	1.180	0.697	15.910	10.230	153.90	3.081	85.55	0.527	269.900	910.100
23.2	1.190	0.621	16.590	10.860	163.6	3.267	90.64	0.559	285.640	904.360
24.2	1.200	0.547	17.290	11.500	172.60	3.455	96.07	0.590	302.050	897.950
25.2	1.210	0.473	17.990	12.140	182.10	3.646	101.20	0.623	318.170	891.830
26.00	1.219	0.412	18.590	12.700	190.37	3.811	105.75	0.652	332.290	886.210
26.2	1.220	0.397	19.920	13.910	190.20	4.154	103.70	0.711	332.990	887.010
27.1	1.230	0.312	30.950	21.680	188.60	6.482	92.20	1.171	341.400	888.600
28.1	1.240	0.241	41.130	29.180	188.40	8.716	81.63	1.488	350.790	889.210
28.5	1.245	0.207	46.070	32.840	188.29	9.811	76.48	1.678	355.380	889.620
29.0	1.250	0.175	50.890	36.450	188.40	10.890	71.48	1.863	360.150	889.850

Appendix 7:

Density	Volume	4	4.5	5	5.5	6	6.5	7	7.5	8	8.5	9	9.5	10	10.5	12.5	15	18	20	21	23	25	26.5	28
3.50	1.0000																							
4.00	0.8719	1.0000																						
4.50	0.7823	0.8972	1.0000																					
5.00	0.6926	0.7944	0.8854	1.0000																				
5.50	0.6329	0.7258	0.8090	0.9137	1.0000																			
6.00	0.5731	0.6573	0.7326	0.8275	0.9056	1.0000																		
6.50	0.5304	0.6083	0.6780	0.7658	0.8381	0.9255	1.0000																	
7.00	0.4877	0.5594	0.6235	0.7042	0.7706	0.8510	0.9195	1.0000																
7.50	0.4557	0.5226	0.5825	0.6579	0.7200	0.7951	0.8591	0.9343	1.0000															
8.00	0.4236	0.4858	0.5415	0.6116	0.6694	0.7391	0.7986	0.8686	0.9297	1.0000														
8.50	0.3987	0.4573	0.5097	0.5757	0.6300	0.6957	0.7517	0.8175	0.8750	0.9412	1.0000													
9.00	0.3738	0.4287	0.4779	0.5397	0.5907	0.6522	0.7048	0.7665	0.8204	0.8824	0.9375	1.0000												
9.50	0.3539	0.4059	0.4524	0.5110	0.5592	0.6175	0.6672	0.7257	0.7767	0.8355	0.8876	0.9468	1.0000											
10.00	0.3340	0.3831	0.4270	0.4822	0.5278	0.5828	0.6297	0.6848	0.7330	0.7885	0.8377	0.9438	0.9438	1.0000										
10.50	0.3177	0.3643	0.4061	0.4586	0.5019	0.5543	0.5989	0.6513	0.6971	0.7499	0.7967	0.9510	0.8976	0.9510	1.0000									
11.00	0.3013	0.3456	0.3852	0.4350	0.4761	0.5257	0.5681	0.6178	0.6613	0.7113	0.7557	0.9485	0.8514	0.9021	0.9485									
11.50	0.2878	0.3300	0.3678	0.4155	0.4547	0.5021	0.5425	0.5900	0.6315	0.6793	0.7217	0.9550	0.8131	0.8615	0.9059									
12.00	0.2742	0.3145	0.3505	0.3959	0.4333	0.4785	0.5170	0.5622	0.6018	0.6473	0.6877	0.9529	0.7748	0.8210	0.8632									
12.50	0.2627	0.3013	0.3358	0.3793	0.4151	0.4584	0.4953	0.5387	0.5765	0.6202	0.6589	0.9581	0.7423	0.7865	0.8270	1.0000								
13.00	0.2512	0.2881	0.3211	0.3627	0.3969	0.4383	0.4736	0.5151	0.5513	0.5930	0.6300	0.9562	0.7098	0.7521	0.7908	0.9562								
13.50	0.2414	0.2768	0.3085	0.3485	0.3814	0.4211	0.4550	0.4949	0.5297	0.5698	0.6053	0.9608	0.6820	0.7226	0.7598	0.9187								
14.00	0.2315	0.2655	0.2959	0.3342	0.3658	0.4039	0.4365	0.4747	0.5081	0.5465	0.5806	0.9592	0.6541	0.6931	0.7288	0.8812								
14.50	0.2230	0.2557	0.2850	0.3219	0.3523	0.3890	0.4203	0.4571	0.4893	0.5263	0.5592	0.9631	0.6300	0.6675	0.7019	0.8487								
15.00	0.2144	0.2459	0.2741	0.3096	0.3388	0.3741	0.4042	0.4396	0.4705	0.5061	0.5377	0.9617	0.6058	0.6419	0.6750	0.8161	1.0000							
15.50	0.2070	0.2374	0.2646	0.2988	0.3270	0.3611	0.3902	0.4243	0.4542	0.4886	0.5191	0.9653	0.5848	0.6196	0.6515	0.7878	0.9653							
16.00	0.1995	0.2288	0.2550	0.2880	0.3152	0.3481	0.3761	0.4091	0.4378	0.4710	0.5004	0.9640	0.5637	0.5973	0.6280	0.7594	0.9305							
16.50	0.1929	0.2212	0.2466	0.2785	0.3048	0.3366	0.3637	0.3955	0.4234	0.4554	0.4838	0.9669	0.5451	0.5775	0.6073	0.7343	0.8997							
17.00	0.1863	0.2137	0.2382	0.2690	0.2944	0.3251	0.3512	0.3820	0.4089	0.4398	0.4673	0.9658	0.5264	0.5578	0.5865	0.7092	0.8689							
17.50	0.1805	0.2070	0.2307	0.2605	0.2851	0.3149	0.3402	0.3700	0.3960	0.4260	0.4526	0.9686	0.5099	0.5403	0.5681	0.6869	0.8417							
18.00	0.1746	0.2003	0.2232	0.2521	0.2759	0.3047	0.3292	0.3580	0.3832	0.4122	0.4379	0.9676	0.4934	0.5228	0.5497	0.6646	0.8144	1.0000						
18.50	0.1694	0.1942	0.2165	0.2445	0.2676	0.2955	0.3193	0.3472	0.3717	0.3998	0.4248	0.9699	0.4785	0.5070	0.5331	0.6447	0.7899	0.9699						
19.00	0.1641	0.1882	0.2098	0.2369	0.2593	0.2863	0.3094	0.3365	0.3601	0.3874	0.4116	0.9690	0.4637	0.4913	0.5166	0.6247	0.7654	0.9399						
19.50	0.1594	0.1828	0.2037	0.2301	0.2518	0.2780	0.3004	0.3267	0.3497	0.3762	0.3997	0.9711	0.4503	0.4771	0.5017	0.6066	0.7432	0.9127						
20.00	0.1546	0.1773	0.1976	0.2232	0.2443	0.2698	0.2915	0.3170	0.3393	0.3650	0.3878	0.9702	0.4368	0.4629	0.4867	0.5885	0.7211	0.8855	1.0000					
20.50	0.1504	0.1724	0.1922	0.2171	0.2376	0.2623	0.2835	0.3083	0.3300	0.3549	0.3771	0.9725	0.4248	0.4501	0.4733	0.5723	0.7013	0.8611	0.9725					
21.00	0.1461	0.1676	0.1868	0.2109	0.2309	0.2549	0.2755	0.2996	0.3206	0.3449	0.3664	0.9717	0.4128	0.4374	0.4599	0.5561	0.6814	0.8368	0.9450	1.0000				
21.50	0.1422	0.1631	0.1818	0.2053	0.2247	0.2481	0.2681	0.2916	0.3121	0.3357	0.3567	0.9733	0.4018	0.4257	0.4477	0.5413	0.6632	0.8144	0.9198	0.9733				
22.00	0.1383	0.1586	0.1768	0.1997	0.2185	0.2413	0.2607	0.2836	0.3035	0.3265	0.3469	0.9726	0.3908	0.4141	0.4354	0.5265	0.6451	0.7921	0.8946	0.9466				
22.50	0.1348	0.1545	0.1723	0.1946	0.2129	0.2351	0.2541	0.2763	0.2957	0.3181	0.3380	0.9743	0.3808	0.4034	0.4242	0.5129	0.6285	0.7718	0.8716	0.9223				
23.00	0.1312	0.1505	0.1677	0.1894	0.2073	0.2289	0.2474	0.2690	0.2879	0.3097	0.3291	0.9737	0.3707	0.3928	0.4130	0.4994	0.6119	0.7514	0.8486	0.8980	1.0000			
23.50	0.1279	0.1467	0.1635	0.1847	0.2021	0.2232	0.2411	0.2623	0.2807	0.3019	0.3208	0.9748	0.3614	0.3829	0.4026	0.4869	0.5965	0.7325	0.8273	0.8754	0.9748			
24.00	0.1246	0.1429	0.1593	0.1799	0.1969	0.2174	0.2349	0.2555	0.2735	0.2941	0.3125	0.9742	0.3521	0.3731	0.3923	0.4743	0.5812	0.7136	0.8060	0.8528	0.9497			
24.50	0.1217	0.1396	0.1556	0.1757	0.1923	0.2124	0.2294	0.2495	0.2671	0.2873	0.3052	0.9767	0.3439	0.3644	0.3831	0.4633	0.5676	0.6970	0.7872	0.8330	0.9276			
25.00	0.1188	0.1363	0.1519	0.1715	0.1877	0.2073	0.2240	0.2436	0.2607	0.2805	0.2980	0.9762	0.3357	0.3557	0.3740	0.4522	0.5541	0.6804	0.7684	0.8131	0.9055	1.0000		
25.50	0.0955	0.1095	0.1220	0.1378	0.1508	0.1666	0.1800	0.1957	0.2095	0.2253	0.2394	0.8035	0.2697	0.2858	0.3005	0.3633	0.4452	0.5467	0.6174	0.6533	0.7275	0.8035		
26.00	0.0721	0.0827	0.0922	0.1041	0.1139	0.1258	0.1359	0.1478	0.1582	0.1702	0.1808	0.7554	0.2037	0.2159	0.2270	0.2745	0.3363	0.4129	0.4664	0.4935	0.5495	0.6069		
26.50	0.0621	0.0712	0.0794	0.0897	0.0981	0.1084	0.1171	0.1273	0.1363	0.1466	0.1558	0.8613	0.1755	0.1859	0.1955	0.2364	0.2896	0.3557	0.4017	0.4251	0.4733	0.5227	1.0000	
27.00	0.0521	0.0598	0.0666	0.0752	0.0823	0.0909	0.0982	0.1068	0.1143	0.1230	0.1307	0.8390	0.1472	0.1560	0.1640	0.1983	0.2430	0.2984	0.3370	0.3566	0.3971	0.4386	0.8390	
27.50	0.0462	0.0530	0.0591	0.0667	0.0730	0.0806	0.0871	0.0947	0.1014	0.1091	0.1159	0.8868	0.1305	0.1383	0.1454	0.1759	0.2155	0.2646	0.2988	0.3162	0.3521	0.3889	0.7440	
28.00	0.0403	0.0462	0.0515	0.0582	0.0637	0.0703	0.0760	0.0826	0.0884	0.0951	0.1011	0.8723	0.1139	0.1207	0.1269	0.1534	0.1880	0.2308	0.2607	0.2758	0.3072	0.3392	0.6490	1.0000
28.50	0.0367	0.0420	0.0469	0.0529	0.0579	0.0640	0.0691	0.0751	0.0804	0.0865	0.0919	0.9094	0.1036	0.1097	0.1154	0.1395	0.1709	0.2099	0.2371	0.2509	0.2793	0.3085	0.5902	0.9094
29.00	0.0330	0.0378	0.0422	0.0476	0.0521	0.0576	0.0622	0.0677	0.0724	0.0779	0.0828	0.9004	0.0932	0.0988	0.1039	0.1256	0.1539	0.1890	0.2135	0.2259	0.2515	0.2778	0.5314	0.8189
29.50	0.0280	0.0321	0.0358	0.0404	0.0442	0.0489	0.0528	0.0574	0.0615	0.0661	0.0702	0.8485	0.0791	0.0838	0.0881	0.1066	0.1306	0.1604	0.1811	0.1916	0.2134	0.2357	0.4509	0.6948
30.00	0.0250	0.0287	0.0320	0.0361	0.0395	0.0436	0.0471	0.0513	0.0549	0.0590	0.0627	0.8929	0.0706	0.0749	0.0787	0.0952	0.1166	0.1432	0.1617	0.1711	0.1905	0.2104	0.4026	0.6203
31.40	0.0220	0.0252	0.0281	0.0318	0.0348	0.0384	0.0415	0.0451	0.0483	0.0519	0.0552	0.8800	0.0622	0.0659	0.0693	0.0837	0.1026	0.1260	0.1423	0.1506	0.1677	0.1852	0.3543	0.5459
34.10	0.0167	0.0192	0.0213	0.0241	0.0264	0.0291	0.0315	0.0342	0.0367	0.0394	0.0419	0.7591	0.0472	0.0500	0.0526	0.0636	0.0779	0.0956	0.1080	0.1143	0.1273	0.1406	0.2689	0.4144

Density Volume Relationship

Appendix 8: Volume – Density

Data Reference Appendix 7

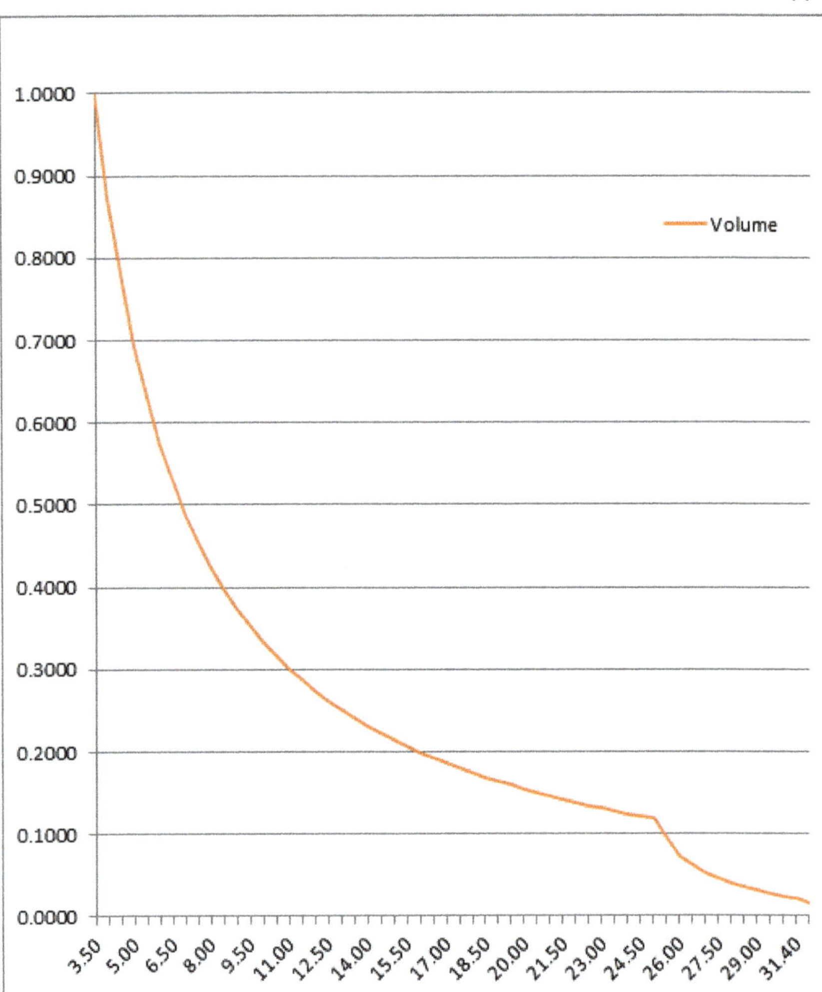

Appendix 9: Hydrometric Table

° Bé	Sp. Gr.	° Bé	Sp. Gr.
0.0	1.0000	20.5	1.1647
0.5	1.0035	21.0	1.1694
1.0	1.0069	21.5	1.1741
1.5	1.0105	22.0	1.1789
2.0	1.0140	22.5	1.1837
2.5	1.0175	23.0	1.1885
3.0	1.0211	23.5	1.1934
3.5	1.0247	24.0	1.1983
4.0	1.0284	24.5	1.2033
4.5	1.0320	25.0	1.2083
5.0	1.0357	25.5	1.2134
5.5	1.0394	26.0	1.2185
6.0	1.0432	26.5	1.2236
6.5	1.0469	27.0	1.2288
7.0	1.0507	27.5	1.2340
7.5	1.0545	28.0	1.2393
8.0	1.0584	28.5	1.2446
8.5	1.0623	29.0	1.2500
9.0	1.0662	29.5	1.2554
9.5	1.0701	30.0	1.2609
10.0	1.0741	30.5	1.2664
10.5	1.0781	31.0	1.2719
11.0	1.0821	31.5	1.2775
11.5	1.0861	32.0	1.2832
12.0	1.0902	32.5	1.2889
12.5	1.0943	33.0	1.2946
13.0	1.0985	33.5	1.3004
13.5	1.1027	34.0	1.3063
14.0	1.1069	34.5	1.3122
14.5	1.1111	35.0	1.3182
15.0	1.1154	35.5	1.3242
15.5	1.1197	36.0	1.3303
16.0	1.1240	36.5	1.3364
16.5	1.1284	37.0	1.3426
17.0	1.1328	37.5	1.3488
17.5	1.1373	38.0	1.3551
18.0	1.1417	38.5	1.3615
18.5	1.1462	39.0	1.3679
19.0	1.1508	39.5	1.3744
19.5	1.1554	40.0	1.3810
20.0	1.1600	40.5	1.3876

Appendix 10: Boiling point of NaCl solution

(in °C)

% of NaCl	Boiling Point	% of NaCl	Boiling Point	% of NaCl	Boiling Point
1	100.21	11	102.66	21	105.81
2	100.42	12	102.94	22	106.16
3	100.64	13	103.23	23	106.52
4	100.87	14	103.53	24	106.89
5	101.10	15	103.83	25	107.27
6	101.34	16	104.14	26	107.65
7	101.59	17	104.46	27	108.04
8	101.85	18	104.79	28	108.43
9	102.11	19	105.12	29	108.83
10	102.38	20	105.46	29.4	108.99

Appendix 11:

Data Reference Appendix 10

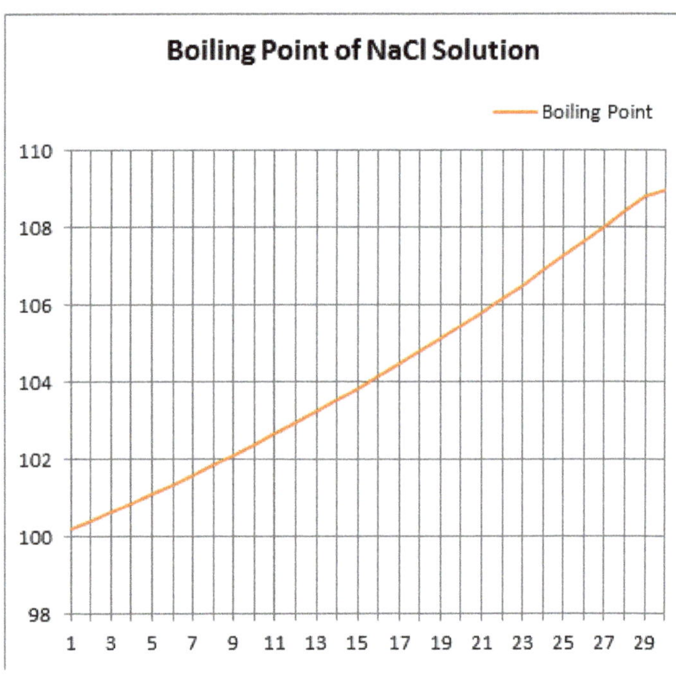

Appendix 12: Solubility of Sodium Chloride in Water

Tempreature °C	Parts by weight of NaCl to 100 H_2O
0	35.63
10	35.69
20	35.82
30	36.03
40	36.32
50	36.67
60	37.06
70	37.51
80	38.00
90	38.52
100	39.12
107.7	39.65
140	42.10
160	43.60
180	44.90

Appendix 13: Solubility of Sodium Chloride in Water

Data Reference Appendix 11

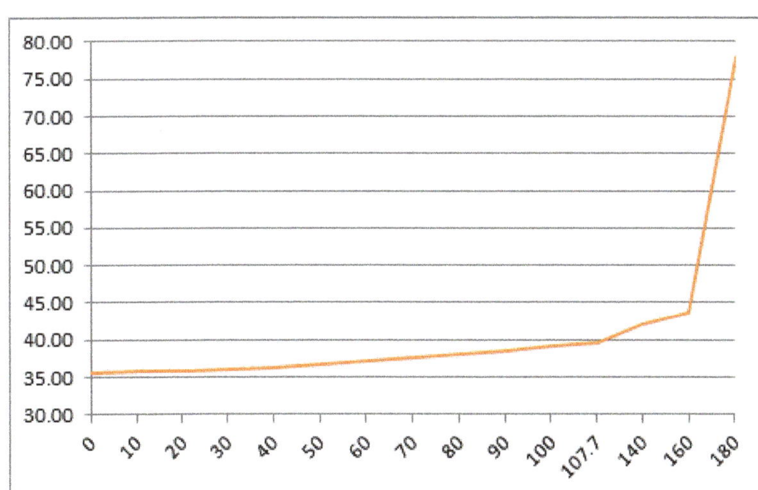

Appendix 14: Solubility of Calcium Carbonate in water at different temperature

°C	gpl of $CaCO_3$
0	0.081
10	0.070
20	0.065
25	0.056
30	0.052
40	0.044
50	0.038

Appendix 15: Solubility of Calcium Sulphate in water at different temperature

	Grams, $CaSO_4$, per 100 grms.	
`° C	Water	Solution
0	0.1762	0.1759
10	0.1932	0.1928
18	0.2020	0.2016
25	0.2040	0.2036
30	0.2094	0.2090
35	0.2100	0.2096
40	0.2101	0.2097
55	0.2013	0.2009
65.3	0.1936	0.1932
75	0.1850	0.1847
100	0.1622	0.1619

Appendix 16: Solubility of Magnesium Sulphate ($MgSO_4$) at Various Temperature

° C	Weight in 100 of water	Weight of water to dissolve 1 part of salt
0	26.9	3.717
5	29.3	3.413
10	31.5	3.174
15	33.8	2.959
20	36.2	2.762
25	38.5	2.623
30	40.9	2.445
35	43.3	2.309
40	45.6	2.193
45	48	2.042
50	50.3	1.988
55	52.7	1.898
60	55	1.818
65	57.3	1.745
70	59.6	1.678
75	61.9	1.615
80	64.2	1.558
85	66.5	1.504
90	68.9	1.451
95	71.4	1.401
100	73.8	1.355

Appendix 17: Solubility of Magnesium Sulphate ($MgSO_4$ $7H_2O$) at Various Temperature

° C	Weight in 100 of water	Weight of water to dissolve 1 part of salt
0	76.9	1.300
10	96.5	1.036
20	119.8	0.836
40	179.5	0.557
70	326.8	0.306
100	671.2	0.149

Appendix 18: Solubility of Potassium Chloride (KCl) at Various Temperature

°C	Weight in 100 parts of water	Weight of water to dissolve1 part of Salt
0	28.5	3.51
5	30.3	3.3
10	32	3.13
15	33.4	2.99
20	34.7	2.88
25	36.1	2.77
30	37.4	2.67
35	38.8	2.58
40	40.1	2.49
45	41.5	2.41
50	42.8	2.33
55	44.2	2.26
60	45.5	2.2
65	46.9	2.13
70	48.3	2.07
75	49.7	2.01
80	51	1.96
85	52.4	1.91
90	53.8	1.86
95	55.2	1.81
100	56.6	1.77

Appendix 19: Solubility of Potassium Sulphate (K_2SO_4) in water at different Temperature

	Grams of K_2SO_4 per 100 grms	
° C	Water	Solution
0	7.35	6.68
1	9.22	8.44
20	11.11	10.00
25	12.04	10.75
30	12.97	11.48
40	14.76	12.86
50	16.5	14.16
60	18.17	15.38
70	19.75	16.49
80	21.40	17.63
90	22.80	18.57
100	24.10	19.42
120	26.50	20.94
143	28.80	22.36
170	32.90	24.76

Appendix 20: Solubility of Sodium Sulphate (Na_2SO_4) in water at different Temperature

	Grams of Na_2SO_4 per 100 grms	
° C	Solution	Water
0	4.76	5.00
5	6.00	6.40
10	8.30	9.00
15	16.3	19.40
20	21.9	28.00
25	29.0	40.80
30	32.8	48.80
40	31.8	46.70
50	31.2	45.30
60	30.4	43.70
80	29.8	42.90
90	29.5	41.95
100	29.6	42.00

Appendix 21: Solubility of Calcium Chloride ($CaCl_2$) in water at different Temperature

	Calcium Chloride $CaCl_2$	
°C	Weight in 100 parts of water.	Weight of water to dissolve 1 part salt
0	49.6	2.016
5	54	1.852
10	60	1.667
15	66	1.515
20	74	1.351
25	82	1.220
30	93	1.075
35	104	0.962
40	110	0.909
45	115	0.869
50	120	0.833
55	125	0.800
60	129	0.775
65	133	0.752
70	136	0.735
75	139	0.719
80	142	0.704
85	144	0.694
90	143	0.680
100	154	0.649

Appendix 22: Solubility of Magnesium Chloride ($MgCl_2$) in water at different Temperature

	Grms $MgCl_2$ in 100 Grm	
°C	Solution	Water
0	34.5	52.8
10	34.9	53.5
25	36.2	56.7
40	36.5	57.5
60	37.9	61
80	39.8	66

°C	Grms $MgCl_2$ in 100 Grm Solution	Water
100	42.2	73
116.7	46.2	85.5
152.6	49.1	96.4
186	56.1	128

Appendix 23: Solubility of Magnesium Chloride ($MgCl_2$ $6H_2O$) in water at different Temperature

°C	Magnesium Chloride $MgCl_2$ x $6H_2O$ Weight in 100 parts of water	Weight of water of dissolve 1 part of salt
25	364.7	0.27
40	416.5	0.24
60	485.6	0.2
80	558.6	0.17

Appendix 24: Solubility of Potassium Bromide (KBr) in water at different Temperature

°C	Grms of KBr per 100 grms Solution	Water
0	34.9	53.5
5	36.1	56.6
10	37.3	59.5
15	38.5	62.5
20	39.5	65.2
25	40.4	67.5
30	41.4	70.6
40	43	75.5
50	44.5	80.2
60	46.1	85.5
70	47.4	90
80	48.7	95
90	49.8	99.2
100	51	104.0
110	52.3	109.5
140	54.7	120.9
181	59.3	145.6

Appendix 24: Solubility of Magnesium Bromide (MgBr) in water at different Temperature

	Grms of MgBr per 100 grms	
°C	Solution	Water
-10	47.2	90.5
0	47.9	91.9
10	48.6	94.5
18	49	96.1
20	49.1	96.5
25	49.4	97.6
30	49.8	99.2
40	50.4	101.6
50	51	104.1
60	51.8	107.5
80	53.2	113.7
100	54.6	120.2
120	56	127.5
140	58	138.1
160	62	163.1

Appendix 25: Vapor pressure of Water

°C	mm	°C	mm
1	4.940	22	19.659
2	5.302	23	20.888
3	5.687	24	22.184
4	6.097	25	23.550
5	6.534	26	24.988
6	6.998	27	26.505
7	7.492	28	28.101
8	8.017	29	29.782
9	8.574	30	31.543
10	9.165	31	33.405
11	9.792	32	35.359
12	10.457	33	37.410
13	11.908	34	39.565
14	12.966	35	41.827
15	13.536	36	44.200

°C	mm	°C	mm
16	14.421	37	46.690
17	15.357	38	49.300
18	15.357	39	52.040
19	16.346	40	54.910
20	17.391	45	71.390
21	18.495		

Appendix 26: Vapor pressure of saturated salt solution

°C	mm	°C	mm
10	7.84	21	14.80
11	7.93	22	15.75
12	8.38	23	16.70
13	8.95	24	17.74
14	9.55	25	18.84
15	10.18	26	19.99
16	10.86	27	21.21
17	11.52	28	22.48
18	12.30	29	23.83
19	13.10	30	25.25
20	13.93		

www.ingramcontent.com/pod-product-compliance
Ingram Content Group UK Ltd.
Pitfield, Milton Keynes, MK11 3LW, UK
UKHW020141300726
14059UKWH00007B/78

9 789388 173070